P9-CKE-168

The Biology of Populations

THE BIOLOGY OF CELLS
    Herbert Stern and David L. Nanney

THE BIOLOGY OF ORGANISMS
    William H. Telfer and Donald Kennedy

THE BIOLOGY OF POPULATIONS
    Robert H. MacArthur and Joseph H. Connell

THE BIOLOGY
OF POPULATIONS

ROBERT H. MacARTHUR   Princeton University

JOSEPH H. CONNELL   University of California
Santa Barbara

John Wiley & Sons, Inc., New York • London • Sydney

LIBRARY

OCT 7 1968

UNIVERSITY OF THE PACIFIC

190462

SECOND PRINTING, MARCH, 1967

Copyright © 1966 by John Wiley & Sons, Inc.

All Rights Reserved
This book or any part thereof
must not be reproduced in any form
without the written permission of the publisher.

Library of Congress Catalog Card Number: 66-21070
Printed in the United States of America

FOREWORD

The idea for this series of three books was conceived in June 1960 at a meeting sponsored by John Wiley & Sons and involving Professors David Nanney (University of Illinois), Robert MacArthur (University of Pennsylvania), Joseph Gall (Yale University), Peter Ray (University of Michigan), William Van der Kloot (New York University Medical School), and Clifford Grobstein (Stanford University). A year later the three books were outlined, discussed, and interrelated during a second meeting held at the Morris Arboretum of the University of Pennsylvania. Most of the authors participated in this second meeting, which profited from the courtesy and counsel of Professor David Goddard (University of Pennsylvania). A sense of common objective emerged from the two meetings, and it is hoped that the books reflect it.

What was this common objective? In broadest terms it was to demonstrate the conviction that the teaching of biology could be reoriented to convey more effectively and forcefully the intellectual revolution which biology was undergoing. We were delighted to find that we shared not only the conviction but also a vision of how to realize it. It is now clear—but was not then—that this vision was common to many biologists. Revision of biological curricula at the university level is rapid and widespread because teachers (many of whom also are investigators) have almost simultaneously seen how far pedagogy lags behind the swift advance of research.

The vision was embodied in several basic decisions on organization of the three volumes. The first decision was adoption of the "levels" approach. This recognizes that the living world is composed of a hierarchy of organizational patterns that can best be encompassed by explicit discussions of cells, organisms, and populations. In parallel with the various levels of organization treated in the physical sciences, each level has its own set of characteristics. These compel the biologist to be concerned with the relation of properties and behavior at one level to each of the other levels. We believe (unlike the vitalists of an earlier era) that much of the behavior of living things is referable to the properties of their components—although we also recognize that many of the properties of the components cannot be ascertained in isolated systems alone. We recognize, also, that biological entities at lower levels of organization are markedly dependent upon the properties of those at higher levels of which they are part. The properties of biological systems have developed from both molecular mechanisms and evolutionary forces; these opposing poles account for much of the pattern and unity of life.

A second decision was to emphasize the uniformities of nature as opposed to its variety. The great generalizations of biology are those that apply to life in all (or many) of its manifestations. The endless diversity of life is a fact, but its endless documentation is inappropriate as an introduction where the major focus should be on life *in toto*.

A third decision concerned assumptions about the previous background of the student. How much information and experience relevant to biology are students entering college likely to have? Improvements in secondary school preparation have been dramatic, not only in biology but also in mathematics and the physical sciences. The prospect is that we shall see increasing numbers of well-prepared students for whom most past college introductions to science are inappropriate and obsolete. Accepting the near-impossibility of accommodating all levels of prior training, we need deliberate attempts to build upon the enriched experience of these students who have already been effectively exposed to some of the basic concepts of physics, chemistry, mathematics, and biology.

Finally, we decided that—although science is frequently regarded as a "body of knowledge," and presented in an up-to-date summary of that body—any such treatment is doomed to early obsolescence. Science is more than a body of knowledge; it also is a process. In the long run, comprehension of the process is more significant than knowledge of any of its particular products. Therefore, it is important to present the intellectual roots of as many topics as possible and the elements essential to their growth.

These were the decisions and the intentions. Their validity, and the effectiveness of our execution, must now be judged in practice. Each of these books may be used alone as a presentation of its particular level. Each, however, has been written under the explicit assumption that no one level of biological organization is fully comprehended without consideration of the other two. Accordingly, the three books will be most meaningful if used together and in their intended sequence—Cell, Organism, Population. Only in this way is the student likely to benefit from the great continuity and cohesiveness that are the products of the last two decades of biological investigation.

The Authors and Clifford Grobstein, *Series Consultant*

PREFACE

Any aspect of biology that deals with more than one organism is a proper subject of population biology. It differs from the biology of cells and the biology of organisms, of course, primarily in its concern with the changes in numbers and in the genetic composition of populations that accompany changes in time and place. There is, however, one other remarkable difference between the biology of populations and that of cells or organisms; this difference is most easily explained in terms of the different answers that may be given to the same question.

Consider the question: "Why has the viceroy butterfly orange on its wings?" The cell biologist might answer: "Because it has orange pigment." He might also elaborate on the biochemical pathway that led to the orange pigment. The organism biologist would prefer an answer that showed how, during the course of development, cells with orange pigment became widespread on the wings. The population biologist, however, would explain that birds often mistake the orange viceroys for orange monarch butterflies which are distasteful to birds and, therefore, are not eaten by them. Other colors of viceroys would be more likely to succumb to predators and leave no descendants.

None of the answers is more correct, or more fundamental, than the others. The first two, however, are micro-descriptions; that is, they deal with the origin and the arrangements of very small units—eventually molecules—which have to do with the orange color. The population biologist's answer, on the other hand, is not a microdescription but concerns a strategy; it is good strategy for a viceroy butterfly to have orange on its wings so that it can deceive the predators, survive to reproduce, and create more orange viceroys. Furthermore—and this is the essential point—it does not matter to the viceroy what biochemistry or what development produced the resemblance to the distasteful monarch butterfly. Although there are several ways of producing a pigment resembling that on the wings of the monarch butterfly, any one of them is acceptable. We can predict only that the viceroy must look orange in order to be a good mimic. We cannot predict the biochemistry that produces this result. Moreover, unless the different biochemical pathways that are possible also have different side effects of importance, the population biologist is not even interested in knowing what processes are involved.

Experiments performed by population biologists indicate another difference between population biology and other branches of science. Anyone familiar with the clear-cut experiments of classical physicists or with the ingenious experiments of physiologists and embryologists is likely to think that an experiment involves the actual manipulation of objects by the experimenter. The scientist subjects a nerve to some chemical stimulus and records the electrical activity farther along the nerve, or he pinches a young embryo in two and watches how each half develops. Actually, an experiment is only an observation motivated by curiosity. The experimenter says to himself: "I wonder what would happen if the embryo were divided in two halves and each half were allowed to develop independently." The only way for him to find the answer is to divide the embryo into halves himself.

The population biologist, however, often discovers the answer to his question without actively tampering with nature. For example, he may believe that desert mice are pale because that color renders them inconspicuous against their pale sand background. He then says to himself: "I wonder whether these mice would be dark, if their background were dark." He could try to change the background and wait many generations to see whether the mouse color changed. However, it is easier to find a dark lava area in the desert and to verify that in this area the mice are indeed dark. This is a perfectly good experiment! Because the face of the earth offers a wide variety of conditions, the population biologist can frequently find his experiment already performed for him, somewhere. This is why population biologists, on the average, spend more time in the field than in the laboratory.

In addition, populations are more difficult to describe than are cells or organisms. At other levels of organization, a biologist can introduce his subject simply by saying, for example: "Look, here is a cell; it is shaped like this and has these parts, which work as follows." It is very difficult, however, to show someone a population. Populations are immersed in the landscape; they interact with their environment in so many subtle ways that if we try to isolate a population in order to exhibit it we destroy it.

Therefore we begin by describing the setting, both in time and space, in which populations exist. Then we consider how a population developed its properties by the mechanism of natural selection and how different kinds of populations ("species") arose. After we have evolved specific populations, we shall, in the remainder of the book, determine how they function and how they interact with one another. Suggestions for laboratory exercises are appended at the end of the book.

This book is intended to be a rather tough, but strictly

introductory, treatment of population biology. The only biological knowledge assumed is that of genetics (which the student who has read Stern and Nanney, *The Biology of Cells* or its equivalent will already possess). A brief acquaintance with high-school calculus will suffice to enable the student to read the more difficult sections in fine print. Thus the book is intended as part of a demanding introductory biology course. Alternatively, with supplementary reading of articles cited in the text, it should serve as a foundation of a one-semester second-level course in ecology and evolution.

Many people have read parts of the manuscript and have made valuable suggestions. We thank them all. We owe a special debt of gratitude to E. O. Wilson and Monte Lloyd, both of whom read the complete manuscript with great care, making many improvements.

Robert H. MacArthur
Joseph H. Connell

Princeton, New Jersey
Santa Barbara, California
May 1966

CONTENTS

PART ONE

THE PATTERNS

1

Patterns in Time,
Space, and Elsewhere

INTRODUCTION

The distinguished biologist G. E. Hutchinson entitled his recent book
(1965) *The Ecological Theater and the Evolutionary Play*. Part One of
the present book describes the ecological theater in which the evolutionary play is acted. The theater is, of course, the earth in all its
diversity, and the drama has been going on for billions of years.
Because we cannot possibly describe every feature of the earth
or its history, we must select only those aspects that seem fruitful.
This selection is prudent not only for writers who wish their books
to be short but also for scientists who wish to discover generalizations.
There are many unique features of the earth and its history—features
that seem to have occurred in only one place or time. Although these
features are interesting and may, when viewed properly, be useful in
framing generalizations, the *widespread patterns* of space and time are
the most fruitful source of generalizations. Wherever there is a widespread pattern, there is likely to be a general explanation which applies to the whole pattern. In this part of the book we describe the
ecological theater with the goal of finding as many patterns as possible.
The exploration of the mechanisms leading to these patterns occupies
the remainder of the book.

SOME PATTERNS IN TIME: THE CHALLENGE
OF A FLUCTUATING ENVIRONMENT

Unlike Aphrodite, populations do not arrive fully formed on the scene. They grow, develop a structure, move about, and show all the properties of a dynamic equilibrium. Therefore, before we describe their static patterns in space, we shall discuss their dynamic properties.

As we know, the climate is always changing; from minute to minute, from century to century, or over much longer periods, judging from such evidence as fossil forest plants in now barren Antarctica. Although these physical changes are reflected in changes in living populations, the biological ones are almost always less pronounced. From our knowledge of the cell and organism levels of biological organization, we know that within living organisms the amplitude of fluctuation of the outside physical factors is almost always reduced. This is accomplished by various *regulatory* ("homeostatic") mechanisms; classic examples in humans are the control of body temperature and blood chemistry.

The universal occurrence of such regulation in living organisms implies that living processes are better carried on under relatively constant conditions. Any random or cyclic variation in environmental conditions is a challenge that must be countered by an appropriate corrective response of the organism. Thus the amount and variety of regulatory activity required for maintenance of the steady state is dictated by the environment.

The particular regulatory mechanisms used depend on the period of the variation. For example, if the temperature changes slightly from minute to minute as puffs of clouds pass over the sun, a rabbit may respond to the shade with slight shivers or by erection of its hairs. As the temperature falls in late afternoon, the rabbit responds by changes in its activity and, eventually, by seeking shelter from the wind in dense vegetation or a burrow. Some mammals and birds huddle together during the night, thereby decreasing the ratio of their radiating surfaces to their heat-producing volumes. Over longer periods of temperature change, such as summer to winter, these physiological and behavioral mechanisms are insufficient, and others are brought into play. Mobile organisms, such as birds, butterflies, and large mammals and fish, simply leave for warmer places. Less mobile ones may insulate themselves from damage by hibernating in shelters, dropping fragile leaves, or passing the winter in some resistant form such as a seed or spore.

Thus the period of fluctuation determines the choice of regulatory mechanism. However, the period cannot be expressed in absolute terms: it must be related to the life history of the particular organism. To a rotifer whose life-span is 10 days, weekly changes may be as important as are yearly changes to a bird. When the period of change greatly exceeds the life-span of an organism, its regulatory mechanisms probably shift from physiological or behavioral to genetic ones. For example, if the mean annual temperature were to decline over a series of years (as occurred at intervals during the "little ice age" of the sixteenth to the nineteenth centuries), the population would "regulate" by changes in gene frequency. Those offspring whose genotype rendered them more resistant to cold would be able to survive and would themselves leave more offspring. The character of the population is thus altered genetically when the period of fluctuation of the environment is long in relation to the life-span of the organism concerned.

If the change lasted long enough, the population would have changed so much genetically (and, as a consequence, in appearance and behavior) that it probably would be called a new "species." Thus regulation in response to long-period changes is *evolution,* although the short-term physiological responses are genetically determined and also evolve.

Incidentally, population processes may themselves cause changes even in a perfectly constant environment. As populations grow in size, the presence of more individuals modifies the local climate. Beetles in stored flour or grain live in local aggregations, rather than being uniformly scattered. When these populations grow sufficiently dense, their combined muscular activities raise the temperature. In fact, pest-control technicians locate the beetles from these "hot spots" in granaries. Temperatures and humidities also are changed in wild colonies of bees and termites. Forests greatly modify all aspects of the local climate.

It is thus obvious that organisms must counter challenges from both physical and biological sources. Potential competitors and predators offer challenges which are as real as seasonal shifts in rainfall or heat. A baboon regulates just as surely against the challenge of a lion by running up a thorn tree as it regulates by shivering when a cold wind blows. The genes which determine that a holly leaf has thorns were incorporated into the genotype of these plants when the threat of herbivorous animals became serious. Moreover, these biological challenges are all populational ones. An isolated instance of a lion attacking a single baboon is not sufficient for the development of genetically determined traits, such as the ability to climb trees. The continual challenge of a population of lions to baboons, or of grazing animals to holly, is necessary for the evolution of regulatory traits. Populations show a degree of internal cohesion by social behavior; small birds "mob" an owl, and wolves hunt in packs, for a lone wolf cannot subdue a large elk.

This example shows that not all challenges involve changes for the worse. The short growing season at high latitudes is a challenge, which demands that populations take advantage of the abundant food and grow quickly during that time. In utilizing this food they may be challenged, of course, by another population of competitors. Their response may be either to become more efficient in securing the limited resource or else to be more aggressive in defending it against the competitor.

In summary, the period and degree of severity of changes in environment dictate the mechanism of regulatory response of an organism. Minor, short-period fluctuations are countered by physiological or behavioral responses. More severe fluctuations of longer periods are countered by shifts in populations, such as movements, growth or decline in numbers or social behavior. If the fluctuations are so severe that only a small portion of the population can survive them, the survivors pass on the traits that enabled them to withstand the change, and so the genetic character of the population shifts. (The environmental change need not be this drastic; if all survive, but certain individuals leave proportionately more offspring, their genetically determined traits will increase proportionately.)

Regulation over short periods can be studied directly; it is possible but more difficult to study directly changes in the genetic character of populations. Because the rate of change is dependent on the generation time, genetic changes can be studied directly only in small organisms which develop quickly. Thus the population genetics of small insects, protozoans, fungi, and bacteria is better known than that of vertebrates. In contrast, the really long-term changes over millions of years, which are so interesting and valuable for our understanding of biology, are known mainly from the fossils of larger animals and plants.

THE EVIDENCE FROM PAST HISTORY

In the remainder of this book we consider the direct evidence of population changes over relatively short periods of time. At this point we would like to discuss the indirect evidence of changes over long stretches of past time. Because this extends our experience over a time span of about two billion years, it ought to give us additional insight.

One principal difficulty offsets, somewhat, the advantages of a historical viewpoint. This difficulty is the small sample of past life which the existing fossils provide. First, we have found only a tiny fraction of existing fossils. But, more important, think of the small fraction of living things that become fossilized! A dead mammal is virtually always decomposed by carrion-eaters from vultures through beetles to bacteria, and nothing is left to change into a fossil. Perhaps a bone of the carcass gets carried off by a fox and happens to fall into a bog or other place suitable for fossil formation; even then the evidence is nearly worthless, for it tells us almost nothing except that the owner of the bone lived in that particular era. Knowledge of its population density, principal predators and competitors, food—all of these kinds of knowledge are exceedingly difficult to infer, except from the extremely rare areas in which whole communities are preserved as fossils. It turns out (Simpson, 1965) that events taking place more quickly than over about ten thousand years are usually not clear from the fossil records (except perhaps for the last ten thousand years about which our knowledge is more detailed). Thus we have a view of about two billion years of life, but it is a view with poor resolution. Let us first see how this view is attained.

When a paleontologist finds a bit of fossil bone, some charcoal in a cave, some peat in an old lake bottom, or a bit of calcareous shell from an old fossil mollusk, he often makes astounding inferences about how many years old the find is, what forms of life existed together, and even what the climate was like. How does he do this? The knowledge necessary for such statements has built up slowly. Experts gradually inferred that most plants have kept their climate preferences: a fossil forest like present-day tropical rain forests was probably found in a warm, moist region and tundra in a very cold climate. Geologists pointed out that lower layers of rock (except when they had folded) were laid down earlier than upper layers, so that a rough time sequence (without dates) could be established. This kind of prediction has been possible for nearly a century. More recently many new sources of information have been found, and astonishing new inferences have been made. We mention briefly two inferences which are typical of recent advances (Hutchinson, 1953).

When radiation hits the earth's atmosphere, some of the nitrogen is changed into the radioactive isotope of carbon known as carbon 14 (C^{14}) and becomes oxidized to CO_2. This is chemically similar to CO_2 from the usual carbon (C^{12}), so that it is incorporated into plants at a rate proportional to its relative abundance. Radioactive isotopes decay or disintegrate, by the emission of detectable particles, into nonradioactive elements. All molecules of a radioactive substance appear to have, in any given second, equal chances of disintegrating, so that the more radioactive molecules there are, the more are normally disintegrating in any second. Thus, when radioactive carbon is produced at a relatively constant rate, the amount in the atmosphere increases until its rate of decay just balances its constant rate of production. It has then reached an equilibrium. Plants also, as long as they are photosynthesizing, maintain the equilibrium concentration. Because this method gives the correct age for Egyptian artifacts known to be 5000 years old, we conclude that this equilibrium has probably not changed much, and that ancient trees must have incorporated about the same proportion of radioactive C^{14} as do modern ones. This fact enables us to estimate the age of any old piece of wood or charcoal by determining the amount of C^{14} left in it. In fact, about half of the radiocarbon present decays every 5600 years (i.e., its half-life is about 5600 years). This means that the remains of a tree that died 5600 years ago has half the concentration of C^{14} of a modern one. For example, a piece of wood $2 \times 5600 = 11,200$ years old has $\frac{1}{2} \times \frac{1}{2} = \frac{1}{4}$ of a modern piece; a piece $3 \times 5600 = 16,800$ years old has $\frac{1}{2} \times \frac{1}{2} \times \frac{1}{2} = \frac{1}{8}$ of the amount in the present-day wood. Modern means of detecting radioactivity are sufficiently accurate to date bits of wood as old as about 50,000 years with some confidence. Some other radioactive elements, found in rocks, decay less rapidly and can be used for telling the ages of much older rocks, so that the ages of nearly every layer of rocks and associated fossils and other debris can be quite accurately estimated. An important example of the use of radiocarbon dating is the determination of glacial dates: by dating bits of organic matter underlying glacial sediments, we can say that the most recent glaciation took place in the last 30,000 years (Flint, 1957).

Even more impressive is the recent technique of using ratios of two isotopes to ascertain ancient temperatures. The ratio of O^{16} to O^{18} incorporated into carbonates depends on the temperature at which the formation took place. Hence by measuring the ratio in, for example, a fossil shell, we can determine the temperature of the ancient sea at the time of the shell's formation! Emiliani and others have done this with spectacular success, although there are still difficulties in the method that are not resolved.

Finally, we should mention the methods used by paleontologists to reconstruct animals from their fossil remains, although these methods are not a recent discovery. How, for instance, can we determine from examining a skull that the owner of the skull walked erect? Or that he was a tool-user? The tool-using part is relatively simple: if the fossils are repeatedly found associated with bits of old tools, it is at least reasonable to assume that the creatures used them. But how do we know he stood erect? This is a simple deduction from the principles of statics in physics. If an animal's head is to hang in front of him, it needs strong support from the neck to the top of the skull and a good place to which it can be attached (see Fig. 1-1). If the head is to balance atop a vertical spinal column however, it needs only weak support and that support must be symmetrically arranged. Thus a glance at the nuchal area of a fossil skull tells us whether it was atop a vertical spine or in front of a horizontal one. Similar deductions are the concern of paleontologists reconstructing the life of the past from fossil evidence.

Armed with these and other techniques, what have paleontologists learned? First, they have progressed far in reconstructing the ancestry of modern forms, especially of vertebrates. For instance, Simpson (1951) has published a beautiful account of the horse's ancestry and of the complicated tree of horse cousins which branched off and are now often extinct. Figure 1-2 shows the remarkable amount of knowledge we have about horse ancestry. It is not the purpose of this book to summarize these data, any more than to list all of the forms of life now found. We must mention, however, that not every family tree is so well worked out as that of the horses. In fact, many forms—the higher plants and insects, for example—seem to have developed so quickly that no disentangling of ancestry is possible at present.

We know much about our own ancestry and are learning more every year (Le Gros Clark, 1963). No unprejudiced person can fail to be impressed by the fossils of manlike creatures (most of whom walked erect and used tools) which preceded us on this planet. They were not necessarily our ancestors—they may represent side branches which came to naught—but they are very similar to what we would expect our ancestors to be like if we were to assume that we share a common ancestor with the higher apes.

The subject of human ancestry is so fraught with prejudices and preconceptions of one kind or another that we must digress and say a few words about it. To begin with, what about the missing link which is alleged to separate man and the apes? This question is misleading in two ways. First, biologists do not claim that apes are man's ancestors. They assert only that man and apes share a common ancestor more recently than, for example, man and the shrews, which, they say, also share a common ancestor. Thus we need not find a link precisely intermediate between modern man and modern apes. Second, the word

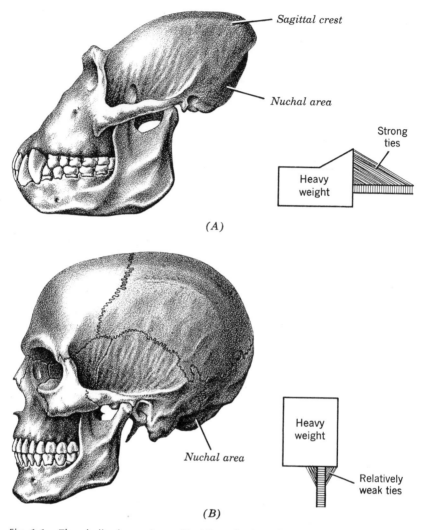

Fig. 1-1. The skull of a male gorilla (A) and of modern *Homo sapiens* (B) seen from the side, to illustrate some of the anatomical landmarks to which reference is made in the text. (Clark, 1963.)

"missing" is misleading. It is inconceivable that we could ever find the fossil remains of every single one of our remote ancestors, and every single one whose remains we fail to find constitutes a missing link. The more complete the fossil record is, the more missing links there will be. We can now proudly state that because we have found many connecting links in man's ancestry, there are now many missing links to fill in!

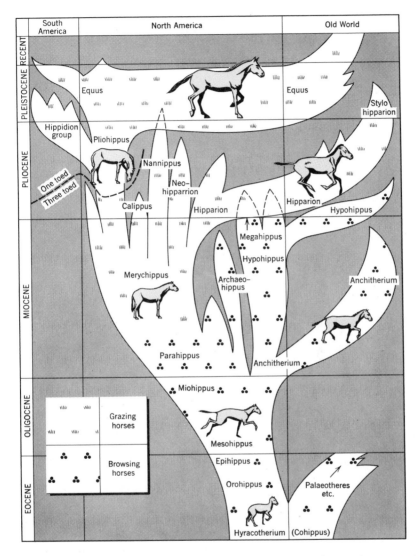

Fig. 1-2. The lineages of the horse family. The main lines of horses descent and relationships of the more important genera. The restorations are to scale. (Simpson, 1951.)

Next, how accurate are the records? What about the Piltdown Hoax? It is true that an amateur archaeologist deceived a keeper of the British Museum and the world at large by cleverly planting fakes of one kind of fossil and another. This hoax, the implications of which (that the brain became manlike before the jaw) were contrary to discoveries in all other fossil finds, slowed down progress for fifty years. How can we be certain that there are not more hoaxes? We cannot, of course, be certain. It is significant, however, that the hoax was exposed by the scientists of the British Museum who did so in the interests of honesty, rather than by someone else with the intention of discrediting all scientific work along these lines. Furthermore, the thoroughness and ingenuity of testing modern fossils make it now virtually impossible to perpetrate another similar hoax which could remain undiscovered for long. There are charlatans in every sphere of life, including science. Science, however, is public knowledge, and becomes discredited as soon as proper contrary evidence is assembled.

What else has paleontology taught us? What principles emerge? Here we can only sample from the many experiments, the record of which has been read by paleontologists. First, let us examine Figs. 1-3, 1-4, and 1-5 which give an estimate of the relative diversity of

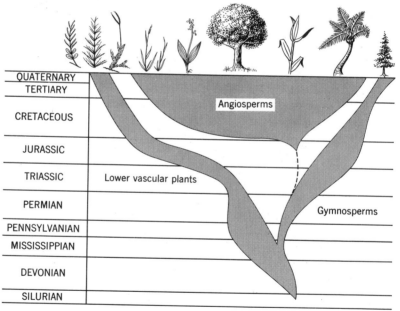

Fig. 1-3. History of land plants shows the spectacular rise of angiosperms in the last 135 million years. The bands are roughly proportional to the number of genera of plants in each group. Angiosperms are flowering plants, a group that includes all the common trees (except conifers), grasses, and vegetables. Lower vascular plants include club mosses, quill worts, and horsetails. The most familiar gymnosperms (naked-seed plants) are the conifers, or evergreens. The diagram is based on one prepared by Erling Dorf of Princeton University. (Newell, 1963.)

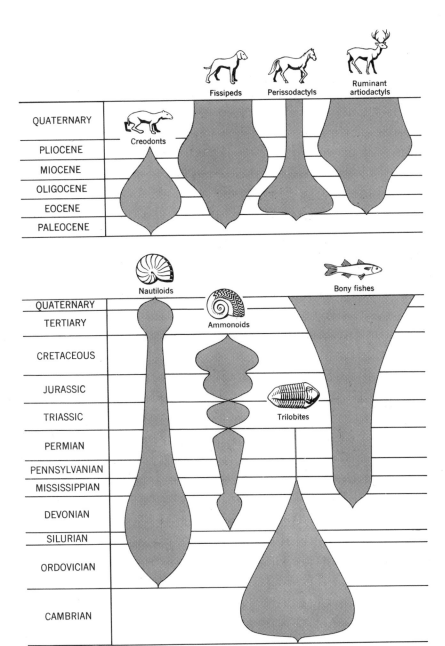

Fig. 1-4. Ecological replacement appears to be a characteristic feature of evolution. The top diagram shows the breadth of family representation among four main groups of mammals over the last 60-odd million years. The bottom diagram shows a similar waxing and waning among four groups of marine swimmers, dating back to the earliest fossil records. The ammonoid group suffered near extinction twice before finally expiring. (Newell, 1963.)

13

living organisms in the course of evolution. Although the number of species has clearly shown a net increase, the remarkable thing is that there were times, particularly the Cretaceous and Tertiary periods, corresponding to the rise of flowering plants and terrestrial vertebrates, when the diversity was expanding very rapidly. At other times, for stretches of hundreds of millions of years, there was little increase at all. This has considerable bearing on the functioning of populations, discussed in later chapters, for it suggests that, except when a wholly new way of life becomes possible, increase in the total number of species is at a comparative standstill. This implies that there probably has been enough time for species to shift about, so that habitats have had equal opportunity for further colonization. Nevertheless, how do we account for the rapid surges that double the diversity of life? In the light of our genetic knowledge (see Chapter 2), we can say that such increases in the diversity of living organisms must have occurred during times when comparatively small proportions of populations survived. The gene substitutions did not merely go in parallel directions, but allowed the organisms to radiate in all directions, to produce a wealth of different phenotypes. This is, of course, just the situation we would expect when colonization of land, with its great diversity of climates and habitats, was at its peak. It is rather like what happened to the diversity of human occupations with the advent of the industrial revolution.

SOME PATTERNS IN SPACE

There are spatial patterns of all sizes. Atmospheric circulation and ocean currents cover distances of many thousands of miles; areas of desert, woodland, and prairie span hundreds of miles. At the other extreme, the spaces between soil particles are measured in microns, and the bacterial environment in which viruses live have even smaller dimensions. In this chapter we begin with large-scale patterns and proceed to the finer ones.

THE GENERAL CIRCULATION OF THE ATMOSPHERE

The climate of the earth, like the life on it, is determined by the energy arriving from the sun. This energy is unequally distributed; at higher latitudes, where the light arrives at a slant, it is spread out over a greater area than near the equator (see Fig. 1-6). Also, the light must pass through a greater amount of the atmosphere when it comes in at

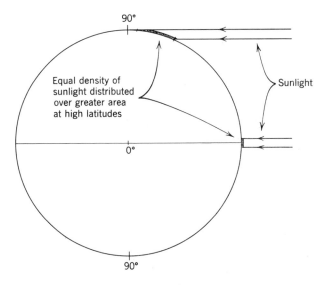

Fig. 1-6

a slant, so that more of the energy is scattered and reflected back into space. Thus less energy arrives per unit area at higher latitudes, and we are not surprised to find lower temperatures at these latitudes.

The radiant energy from the very hot sun is composed of short wavelengths, which are not readily absorbed by water or water vapor in the earth's atmosphere. Therefore, this incident light passes through the atmosphere without much of it being absorbed there; it is either absorbed by the earth's surface or is reflected back to outer space as short wavelengths, by the tops of clouds, snow, and other reflecting surfaces. Because the earth is much cooler than the sun, it reradiates the absorbed energy as longer wavelength "heat waves." These long wavelengths are absorbed by water or water vapor in the atmosphere, so that we can regard the atmosphere as being heated mainly from below. This energy is then reradiated back to the earth and back again in the atmosphere; the moist atmosphere acts as a trap, allowing only a little of this heat to escape to outer space at each exchange. Thus a cold earth would receive more heat than it radiated back; as the earth became warmer, it would radiate more heat and would eventually achieve a balance. This is the present situation. (It is called the "greenhouse effect," in which the water vapor and clouds take the place of the glass.) In addition to this radiation, some of the heat from the earth is transferred to the atmosphere by convection currents of rising air, which carry up heat and evaporated water.

Let us now examine how this unequal heating of the atmosphere from below affects the climate of the earth. We start with an over-simplified explanation, based on a stationary homogeneous earth, and gradually add the complexities of rotation, continents, and seasons.

Because more energy arrives per unit area at the equator, the air, heated from below, expands and, being less dense, rises. Air at the surface on either side of the equator then flows in, replacing the air that has risen. At high elevations the air spills away from the equator. At high altitudes the air is cooled while it moves away from the equator and, therefore tends to fall to the surface near the poles and move toward the equator. Thus a "meridional circulation" is set up; air moves along the meridians of longitude toward the equator at the surface and away from it at high altitudes (see Fig. 1-7a). At the surface the equator is a zone of convergence of air and the pole is a zone of divergence.

However, because of the rotation of the earth, the air leaving the equator does not travel all the way to the poles before sinking. It tends to cool and sink at about latitudes 30° North and South of the equator and follows the pattern shown in Fig. 1-7b. To understand these phenomena we must consider the effects of a rotating earth.

WINDS AND WATER CURRENTS

At this point we bring in our first complicating factor—the rotation of the earth. It has the effect of causing moving objects (and winds) to veer to the right in the northern hemisphere and to the left in the southern hemisphere. This is called the "coriolis force." To under-stand it, imagine an object in the northern hemisphere moving north from some tropical location. Because the earth is roughly spherical, the object, as it moves north, is also coming closer to the earth's axis of rotation; that is, it is moving to a region in which the earth's spin-ning surface is not moving so fast. Because, according to the laws of motion, an object maintains its momentum unless it is acted on by an outside force, such as friction, our object must speed up rela-tive to the earth's rotation (Fig. 1-8). This would cause it to veer in the direction of the earth's rotation (to the east); hence it must veer right. (Satisfy yourself by similar reasoning that a southward moving object would also move to the right in the northern hemisphere and to the left in the southern hemisphere. Also, it should be clear

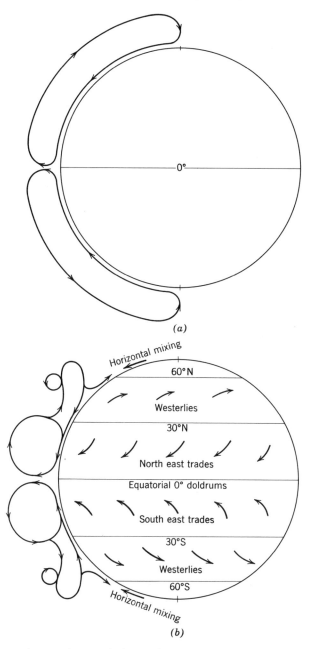

Fig. 1-7. The circulation of the earth's atmosphere: (a) on a hypothetical non-rotating earth, (b) incorporating the effects of the earth's rotation. (After Byers, 1954.)

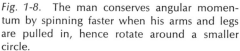

Fig. 1-8. The man conserves angular momentum by spinning faster when his arms and legs are pulled in, hence rotate around a smaller circle.

Fig. 1-9. Same as Fig. 1-8 when rotation is increased.

that if an object moves 100 miles north, beginning at the equator, it comes only slightly closer to the earth's axis (about one mile closer). If it starts at 30°N. latitude, however, it moves about 50 miles closer to the axis by moving 100 miles north along the surface; in fact, the change varies as the sine of the latitude. Because the coriolis force depends on a change in distance from the earth's axis, it is zero at the equator and maximum at the poles.

In Fig. 1-7*b* we see that, at the earth's surface, the meridional circulation moves toward the equator between 0 and about 30°. As these flows veer to the right and left in the northern and southern hemisphere, they acquire an easterly component. This accounts for the northeast and southeast "trade" winds, as shown in the figure. Air flowing away from both poles veers into polar easterly winds. At middle latitudes air moving toward the poles veers into westerly winds.

The force of the winds on the surface of the oceans, coupled with "downhill" flow from areas of higher water level, produces currents. The prevailing northeast and southeast trade winds set up surface currents flowing west above and below the equator. These "equatorial" currents are deflected toward higher latitudes when they strike the east shores of the continents of Asia and America. The water then returns east across the oceans at higher latitudes and flows toward the equator along the western shores of the continents. Thus in both the Atlantic and the Pacific two huge cells of circulating surface water are produced, as shown in Fig. 1-10. They tend to carry heat to higher latitudes in the western part of the oceans, with colder water converging toward low latitudes in the eastern oceans. Some of the water that is piled up on the eastern shores of the continents, instead of being diverted north or south, flows directly back between the two equatorial currents as an "equatorial countercurrent."

THE GENERAL PATTERN OF MOISTURE

As a result of the general pattern of circulation of air, an unequal distribution of moisture is produced in the following manner.

First, cold air holds less moisture, in gaseous form, than warm air (everyone has seen the water vapor in warm air condense when it strikes a cold object, such as a glass of ice water). Second, as air rises, it expands, because it then is under less pressure (less weight of air above). Third, a gas does work when it expands, hence uses energy. If it expands without taking this energy from an outside source (which is what happens when air rises without contact with a surface), the energy must come from within the gas itself, that is, from its own heat energy. Then the gas is cooled "adiabatically." Thus when air rises, it is cooled (in the same way that air is cooled as it expands when released from an automobile tire). If warm air containing abundant water vapor rises and so cools, some of the water condenses as droplets to form clouds and, eventually, rain. Conversely, as air falls, it is warmed adiabatically, and any liquid water is changed into water vapor.

Thus, in accordance with the general pattern of circulation, the trade winds moving toward the equator are warmed and pick up water from the surface. At the equator air rises, cools, and produces heavy rainfall. Having lost much of its moisture, the air moves toward 30° latitude, falls, and, as it is warmed, picks up moisture. Therefore there are deserts near 30° latitude, especially on the western sides of continents, where the cold ocean currents are offshore. The cold oceans are covered by a blanket of cold air which keeps the decending air from picking up moisture. When the cold and warm air meet, fog is produced. This fog is the principal source of moisture for plants in these regions.

From 40 to 60° latitudes on west coasts, where westerly winds blow in off a warm ocean, there is another zone of heavy rainfall. There are both local and seasonal variations to this pattern, but the world map of rainfall shown in Fig. 1-11 confirms the general nature of this distribution of rain.

Fig. 1-10. Pattern of the world's ocean currents.

20

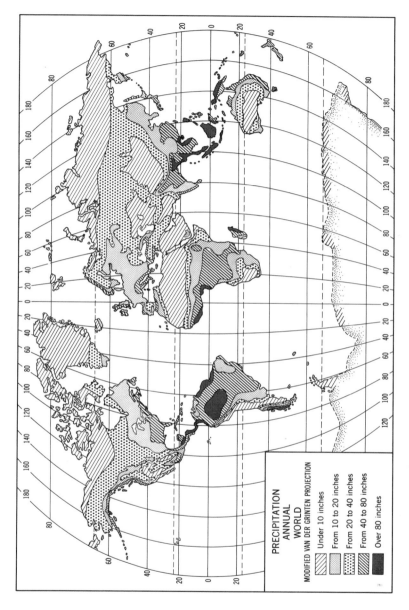

Fig. 1-11. Pattern of world precipitation. (After Koeppe, 1958.)

PRECIPITATION
ANNUAL
WORLD
MODIFIED VAN DER GRINTEN PROJECTION

Under 10 inches
From 10 to 20 inches
From 20 to 40 inches
From 40 to 80 inches
Over 80 inches

VARIATIONS OF CLIMATE FROM PLACE TO PLACE

This general pattern would hold on a homogeneous earth, but the presence of land masses, water bodies, and mountains causes variations. We have already seen one example in the west coast maritime fogs.

The radiant energy of the sun does not penetrate as far into the surface of the land as it does into water; also, some of the heat is carried by convection currents down into water, but not into the land crust. Furthermore, the amount of radiant heat that can raise the temperature of a cubic foot of water 1°F, can raise the temperature of a cubic foot of rock by 4°F and more quickly. Thus the land heats up faster and to a much higher temperature than the water, and it also cools down faster. This difference is sufficiently great to cause daily sea and land breezes on a coastline.

On a vaster scale, such differences in land and sea produce the monsoons of Asia. During the summer the interior of the continent is considerably heated; the dry air of the interior rises, and warm moist air is drawn in from the Indian and Pacific Oceans, bringing rain to the southern and eastern regions. A dramatic example is Cherrapunji, India which received one hundred and fifty inches of rain in five days in August 1841. In winter the interior cools to very low temperatures, the winds are reversed, and cold dry air streams out from the interior over a huge bordering region from Kamchatka to Arabia. Thus "continental" regions have great daily and seasonal changes in temperature and moisture.

In contrast, coastal regions or islands have a more even climate because of the great capacity of water to absorb and hold heat. The heat gained in summer is slowly released, tempering the winter cold; the reverse happens in summer, so that the annual changes in temperature here are smaller than in the interior of continents. In temperate latitudes with westerly circulation the east coasts are more affected by the continental climates than the west coasts.

Mountain chains, which extend into the upper atmosphere, block the movements of air. The climate of much of Canada and the United States is "continental" because the high mountains near the west coasts prevent the flow of "maritime" air into the interior from the Pacific Ocean. No such barrier exists in western Europe, and the maritime influences extend far inland.

Another influence of mountains is called the "rainshadow" effect. As air is driven against a mountain, it is forced upward and, as it expands and cools adiabatically, its moisture is condensed and lost in rain. On the lee side of the mountain the air falls, is warmed, and picks up moisture. Thus the land downwind of a mountain is in its "rainshadow." The Caribbean islands provide dramatic examples. The northeast trades lose their water in the central mountains and leave rainshadow deserts in the southwest.

Because air cools as it rises, the average temperature decreases with increasing altitude. Thus a trip up a mountain reproduces in miniature a trip toward higher latitudes. The vegetation of a tropical mountain may change from rain forest at the base to tundra or ice at the top if the mountain is sufficiently high.

Conversely, if we descend into deep water, the temperature does not change much, but the light quickly fades, different wavelengths penetrating different distances. The pressure increases rapidly; it is difficult to separate the effects of pressure from those of the absence of light. Thus, as increased altitude is accompanied by both decreased pressure and temperature, increase in depth is accompanied by increased pressure and decreased light.

SEASONAL VARIATIONS

The seasonal march of the sun, relative to the earth—from directly over 23°S. latitude about December 21st to directly over 23°N. latitude about June 21st—causes the earth's seasons. Imagine the illuminated half of the globe on some day when the sun is north of the equator (northern summer), as shown in Fig. 1-12. The north pole will inevitably be illuminated (hence the six months' summer at the pole), and a part larger than half of any parallel of latitude north of the equator will be illuminated (hence longer days in summer). By symmetry, the excess of night at the same parallel of latitude south of the equator will just equal the excess of day north of the equator. This is important, for, as the sun's relative position moves north and south symmetrically, we can say that the excess of day at one time in summer precisely equals the deficiency of day (or excess of night) at the same place at the corresponding time in winter; that is, if we add the daily duration of daylight of a day x days after June 21st and that of a day x days after December 21st, we will always obtain twenty-four hours. Thus the total summer excess just balances the total winter deficit. This holds true for any latitude, so that we can say that the total amount of daylight during the whole year is just six months at *every latitude*. The poles receive their six months of daylight in one piece, the equator has twelve hours of daylight all year long, and intermediate places have longer days in summer and shorter days in winter; all areas, however, receive the same total amount of daylight.

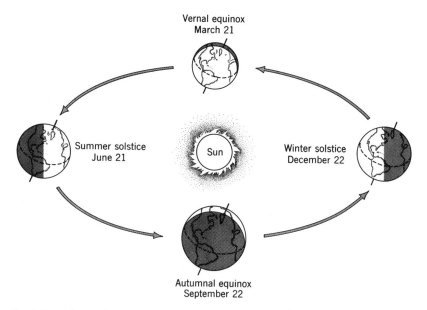

Fig. 1-12. The sunlit portions of the northern hemisphere are seen to vary from greater than one half in summer to less than one half in winter. The proportion of any latitude that is sunlit is also the proportion of the twenty-four hour day between sunrise and sunset.

Where the temperature does not change much, as in the sea, seasonal changes in light may have considerable effects. In the North Atlantic the increase in light in spring produces a vast acceleration in growth of plants, a "spring bloom," with almost no accompanying change in temperature.

Aside from the many obvious effects of seasonal changes in temperature on land communities, there are interesting indirect effects on water bodies. The surface water of a deep lake or ocean is heated in summer, and the warm water, because it is lighter, remains at the surface. A layer of warm water is formed above the deeper, denser cold water, with which it does not mix. The deep water is thus isolated from the atmosphere and may become depleted of oxygen; the surface layer, isolated from the bottom, may become depleted of mineral nutrients. As the surface cools in the autumn, a point may be reached when the whole lake is at about the same temperature. At this point, wind stresses on the surface cause a "turnover" of the water, thereby mixing the two layers and renewing the mineral nutrients at the surface and the oxygen content of the depths.

Vertical mixing of water is also accomplished when surface water is pushed away from a shore; it is then replaced by the "upwelling" of deeper water. This occurs with prevailing offshore winds or when a current running along a coast veers away due to coriolis force. As can be seen in Fig. 1-10, the main ocean currents move along the west coasts of the continents in the direction that would tend to cause the surface water to veer offshore and produce upwelling. In these regions this brings mineral nutrients up to the surface; they are regions of high productivity.

One other seasonal difference involves the interaction of temperature and moisture. Molecules of water continually escape from the surface into the air, but more of them escape per unit time when they are more active, as they are at higher temperatures. Thus evaporation increases with temperature. With the same amount of rainfall, a higher temperature results in greater evaporation, and therefore less water remains for the plants. For example, in northern Australia the rains come mainly in summer; the availability of this rainfall on plant growth is thus decreased. In southern Australia, which has winter rains, the decreased evaporation then means that a larger part of the rainfall is available for plant growth. The availability of the rainfall is a function of the temperature during the rainy season.

If the rains are strictly seasonal, as in much of Australia, the *difference* between the seasons is greatest when rains come in winter. The summer is extra dry because the evaporation is greatest during that time. With summer rains the *difference* between the moisture supply in summer and winter is decreased (Fig. 1-13).

THE PATTERNS OF LAND ORGANISMS

We are now in a position to predict very crudely the climate of a land region from knowledge of its latitude, elevation, and position in a continent. The next question is: can we now predict the kinds of life that will be found there? A little reflection shows that we cannot expect great accuracy in the prediction: different soil types favor different kinds of plants, different kinds of insects in turn depend on these plants, and so on. Furthermore, diverse histories of forest cutting, agriculture, etc., will have left their marks. To a remarkable extent, however, knowledge of temperature and of the amount and seasonal pattern of rainfall allows us to make some general predictions. Perhaps the most direct approach to the widespread pattern of the world's vegetation is due to L. R. Holdridge (1947), who has constructed the

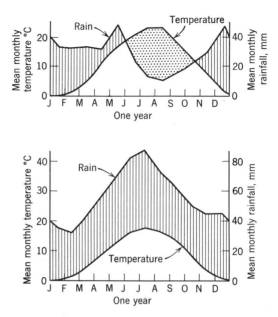

Fig. 1-13. The relation between rainfall and temperature in two localities with very different climatic conditions. *Above:* Ankara, Turkey—drought during summer; *below:* Hohenheim, Germany—no drought stresses. (After Walter and Lieth, 1960.)

triangle shown in Fig. 1-14. In many ways this is the best scheme yet proposed; it is certainly the simplest. Of course, its very simplicity leads to some inaccuracies. We showed earlier how the excess of precipitation over evaporation might be found principally in summer in some areas of the earth and in winter in other areas. Holdridge's scheme sums the temperature and rainfall over the whole year, hence does not explicitly take the seasons into account. In spite of this kind of difficulty, it is nevertheless surprisingly good in its prediction of the "big picture" of vegetation across the world. Needless to say, this prediction applies only to areas that are relatively undisturbed by man; it does not attempt to predict what crops man will grow in agricultural areas!

Notice that Holdridge's scheme is predictive; many ecologists prefer to be less daring and simply describe the vegetation pattern and its associated animals; that is, instead of relating vegetation to climate, they relate it to geography, and instead of making a prediction of the kind that Holdridge attempted they construct an empirical map of a continent or the world with vegetation types, as they actually exist, drawn on it. We include an example of a vegetation map of the world constructed by Odum (1959) (Fig. 1-15).

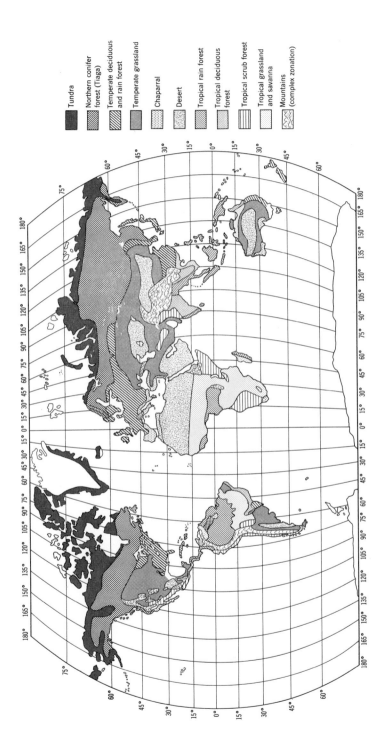

Fig. 1-15. Schematic map of the major vegetation types of the world. (After Odum, 1959.)

Tundra

Northern conifer forest (Tiaga)

Temperate deciduous and rain forest

Temperate grassland

Chaparral

Desert

Tropical rain forest

Tropical deciduous forest

Tropical scrub forest

Tropical grassland and savanna

Mountains (complex zonation)

We now describe, with illustrations (Fig. 1-16 through Fig. 1-23), some of these vegetation types, but we emphasize that they are not usually discrete vegetation types. They are rather more like colors, which we name separately for convenience (red, orange, yellow, etc.), but which, we know, grade continuously into one another. Near the equator we find tropical rain forests. (There are exceptions to this on high mountains and in east Africa, where the deserts to the north influence the patterns and where there is a long history of human disturbance.) On either side of this belt of forests the trees become more widely spaced and shorter: woodlands and savannahs succeed each other into deserts of low shrubs. Rain forests may occur wherever local conditions produce heavy rainfall.

Further from the equator the mean temperature falls and, as the rainfall increases again, deciduous forests appear. In the northern hemisphere they change to needle-leaved evergreen "boreal" forests in Canada, northern Europe, and Asia; as we approach the north pole, trees disappear and only low tundra occurs. This pattern, on a smaller

Fig. 1-16. Mixed hardwood stand in Duke Forest, Duke University, Durham, North Carolina. (Courtesy of American Forest Products Industries, Inc., Washington, D.C.)

Fig. 1-17. Succulent desert, Arizona. (Courtesy of A. Devaney, Inc., New York.)

scale, is repeated in the southern hemisphere; however, there is little land between 45° and 70°S. latitude. In the zone in which northern United States, Canada, Alaska, northern Europe, and Siberia lie, there is only the southern tip of South America and New Zealand in the southern hemisphere. Antarctica, with its vast area of perpetual ice and snow, reflects most of the sun that reaches it, and the only photosynthesizers are lichens and algae.

The fact that we pass through boreal forests and then tundra as we ascend a high mountain in temperate latitudes indicates that lower temperatures are probably the main reason for these vegetative types. In tundra, arctic or alpine, the soil is frozen in winter and thaws only to a shallow depth in summer. This has spectacular effects on human life, for in these regions neither sewer nor water pipes can be buried below the frost line. The permanently frozen subsoil is impenetrable to roots and also prevents drainage; both conditions prevent the growth of trees. Also, plants sufficiently tall to project through the snow cover would be killed during the long arctic winter. On tropical high mountains, instead of the arctic sequence of long, dark winters and short, light summers with low annual rainfall, all days are of about the same length, and the rainfall may be very great. The "boreal" forests are enveloped in clouds most of the time, and the short trees are festooned with epiphytic plants. Higher on the mountain strong winds tend to dry out the upper twigs of trees, and they grow along the ground in a dense mat. Even higher, a tundra-like zone of low perennial herbs and shrubs extends up to the permanent snow line.

Trees occur in all of these types of vegetation, unless they are excluded either by some direct or indirect effect of low temperature as in tundra or by lack of water as in deserts. However, one other important example of the exclusion of trees in an intermediate climate are the grasslands. This situation cannot be predicted from a knowledge of climate alone (with the possible exception of certain valleys in high mountains). Additional information about the kind of soil or the history of disturbances is needed.

For example, plains of "Mitchell grass" occur next to sparse woodlands in Australia, where the climate is one of very light seasonal

Fig. 1-18. An Arabian American Oil Company exploration party makes its camp in the sand mountains of the Rub' al-Khali in Saudi Arabia. This is sand desert at its most barren. (Courtesy of Standard Oil Company, New Jersey.)

Fig. 1-19. A second-growth tropical forest in New Guinea. (Courtesy of Standard Oil Company, New Jersey.)

rainfall. The trees are on the porous sandy soils and the grasslands on the heavily textured fine soils. The grasses with their numerous fine roots can evidently remove all water before it has time to percolate to deeper layers, thereby excluding the tree seedlings which have coarser and deeper roots.

On the eastern edge of the prairies of North America, rainfall is heavier, and no sharp differences in either soil or climate separate the grasslands from the eastern deciduous forests. Here prediction depends on knowledge of the different methods of growth of grasses and trees. Grasses grow from the base of the stem, whereas most trees grow at the tips of the stem and branches. Anything that removes the tops depresses the growth of trees, but not of grasses; thus grasses are favored.

On the eastern prairies huge herds of bison grazed before the Europeans arrived, and fires burned across vast areas in the autumn when the grass was dry. The only firebreaks were rivers and sharp rocky escarpments, and trees occurred along these places. Sauer (1952), reminiscing about his childhood in Missouri, says: "From grandparents I heard of the early days when people dared not build their houses beyond the shelter of the wooded slopes, until the plow stopped the autumnal prairie fires." Trees have invaded the eastern edges of the prairies only since they have had protection from fires; it seems likely, therefore, that before the arrival of Europeans, grazing and annual fires contributed to the existence of a treeless prairie.

On the western edges of the great plains the rainfall is much lighter, and a situation somewhat similar to that in Australia exists. Trees are

Fig. 1-20. A mature tropical forest, Klamono, New Guinea. (Courtesy of Standard Oil Company, New Jersey.)

Fig. 1-21. Tundra and reindeer moss, N.W.T., Canada. (Courtesy of Standard Oil Company, New Jersey.)

on the porous soil of rocky outcrops and grass on the fine soils that have been washed and blown over the level areas. Because some invasion of trees onto the level areas of plains has occurred since the removal of both bison and the annual fires, they may have excluded the trees in earlier times. This is certainly the case in the Florida Everglades, where regular burn-offs are necessary to prevent encroachment of trees into the saw grass. Since the establishment of the National Park, fire control has been nearly perfect, and a large part of the original saw-grass flats is being taken over by trees. This is a case in which conservation requires that we stop putting out fires! Whatever the reason, it seems clear that this kind of vegetation, and probably many others, cannot be predicted solely from a knowledge of climate.

The various communities with different types of vegetation are illustrated in Figs. 1-16 to 1-23. These photographs can also serve as a substitute, although an inadequate one, for visiting these regions. One of the main compensations of a population biologist is the ability, as part of his job, to go and live in many such places. Darwin, accustomed to the English countryside, described his first day in the wet tropical forests in Brazil (Darwin, 1860) as follows:

The day has passed delightfully. Delight itself, however, is a weak term to express the feelings of a naturalist who, for the first time, has wandered by himself in a Brazilian forest. The elegance of the grasses, the novelty of the parasitical plants, the beauty of the flowers, the glossy green of the foliage, but above all the general luxuriance of the vegetation filled me with admiration. A most paradoxical mixture of sound and silence pervades the shady parts of the wood. The noise from the insects is so loud, that it may be heard even in a vessel anchored hundreds of yards from the shore; yet within the recesses of the forest a universal silence appears to reign. To a person fond of natural history, such a day as this brings with it a deeper pleasure than he can ever hope to experience again.

Once we know the general type of land vegetation, we can predict the type of animals found there. We can say that there will be swift, grazing mammals in grassland, but we cannot predict whether they will be the pronghorns of North America, the antelopes of Africa, or the kangaroos of Australia. Their swiftness is an indication that they evolved together with swift predators, such as hunting dogs, wolves,

Fig. 1-22. Tundra grass and reindeer moss, Mackenzie mountain region, Canada. (Courtesy of Standard Oil Company, New Jersey.)

Fig. 1-23. Levittown, Pennsylvania. This is rapidly becoming the commonest habitat. (Courtesy of Philip Gendreau.)

or cheetahs. In contrast, herbivores, such as deer in forests, must be agile rather than merely swift, for predators may attack them from ambush.

DETAILED ACCOUNT OF TROPICAL FOREST

We have so far discussed the large-scale patterns in space. To introduce the smaller-scale patterns, we describe a particular area in more detail.

If you were asked to choose the most favorable place on land you would probably name the wet tropics. The concept of favorability is partly based upon the luxuriance of life there, and so smacks of circular reasoning. However the warm, wet, even climate offers fewer physical challenges to existence than the more unstable, rigorous climates at higher latitudes. Partly as a consequence of this favorable climate, and partly because growth and reproduction can continue all year, the tropical rain forest represents the most complex, diverse, and richest example of life on land. For this reason we describe it in more detail. (Incidentally, this is not an easy task, not only because the tropical rain forest is so complex, but also because it is the least studied of all land communities. Most population biologists, who live and work in the temperate zone, find that tropical areas are expensive and difficult to study.) Recalling from our earlier description that tropical areas have more energy (of the kind that plants use) coming in from the sun, we might expect that plant production would be greater in the tropics. Plant production is measured here in units of kilogram calories per square meter per year and refers to the fuel value of the new growth of plants. If, for instance, a square meter of a farmer's field grows three kilograms of some crop each year and each kilogram, when burned, yields 3000 kilogram calories of heat, the production of that field is 9000 kilogram calories per square meter per year. In a forest only new growth and annual leaves constitute production; therefore, to obtain production, we measure the fuel value of the new wood and leaves. The view that there might be a pattern of higher productivity in the tropics is reinforced when we realize that energy is used not only for growth, but also for respiration and regulation.* The energy which is channeled into regulation and respiration is greater where climates are colder and less stable, as in high latitudes. Hence we might expect the tropics not only to have more energy available, but also to use more of it for production of new material.

From this line of reasoning we may predict that the greatest rates of growth and reproduction on land would be found in a tropical rain forest. According to the few available data, this seems to be true. Ovington (1962) also summarized the published data on the rates of decomposition of leaf litter and the weight of chemical nutrients in various woodlands; they are highest in rain forests, which suggests that the rate of cycling of nutrients is also fastest there. In addition, few nutrients are lost in the streams that drain rain forests, even though there is much leaching of the soil during heavy rains. The roots of the trees are abundant and shallow, and they take up the nutrients as soon as they are released by decomposition of the litter.

Because of the high organic productivity and the small number of physical stresses, the challenges offered to an organism by the rain-forest environment are mainly biological in origin. Predators and

---

* See Herbert Stern and David L. Nanney, *The Biology of Cells*, Wiley, 1965; William H. Telfer and Donald Kennedy, *The Biology of Organisms*, Wiley, 1965.

parasites of all kinds are abundant, plants perch on plants, and competition for light among the plants is intense.

In these forests the number of species is astonishing. Although the pattern of many species is found in most organisms, we illustrate it with trees. In an undisturbed rain forest at latitude 17° in North Queensland, Australia, there were 141 species among the 1261 trees over one-inch diameter on an area of 1¼ acres. At latitude 27° in South Queensland there were 88 species among the 1649 trees over one-inch diameter on 1½ acres (Connell, Tracey, and Webb, 1964). This diversity is difficult to comprehend for someone from a temperate country, where such an area of woods may have ten or fifteen species of trees. Another feature of these forests is the lack of a "dominant" species that would account for most of the trees. By summing up the "basal area" (i.e., the cross-sectional area of tree trunks, measured at breast height) of all trees of each species, it is possible to obtain a rough estimate of the relative amount of organic matter contained in each species. In North Queensland the commonest tree accounted for about 9 per cent of the total basal area; in South Queensland, 16 per cent. In a beech-maple forest in southern Michigan, using the same sized trees, the commonest of the ten species of trees was found to represent 51 per cent of the total basal area (Cain, 1935). Each species in such a "mixed" tropical rain forest tends to be widely scattered. We seldom find a grove of a single species; each tree we come to seems to be different. The impression we obtain is an arrangement of great complexity. The masses of vines, epiphytic ferns, and orchids suspended from the canopy add to it. Occasionally we see a glade of light, where a large tree or branch has fallen in a storm. In the dim light under the closed canopy we can walk easily through the forest, but in these openings everything is growing in a thick tangle. Here, or along a road or stream, the canopy slopes down to the ground. In these openings the small trees, which may have grown little until then, begin to grow rapidly, until the canopy is closed again by the few which have grown the fastest. The tropical rain forest thus represents a pattern of more species in the tropics, each species contributing a smaller fraction of the total material.

In this three-dimensional environment the animals move and feed, largely unseen. One of the best studies was made by Harrison (1962), who described the activities of the birds and mammals of rain forests in Malaya and North Queensland. In the canopy, where most of the plant productivity occurs, the insects and vertebrates eat leaves, flowers, and fruit, or they have a mixed diet of plants and insects. When these animals move above or below the canopy, they are caught in flight by birds and bats or are taken from the branches or trunks by other mammals and birds, which range from the ground up to the canopy. Leaves, flowers, fruit, and seeds fall to the ground and are eaten there by various mammals and birds; some of these animals, such as elephants, deer, tapirs and, in Queensland, wallabies, can reach the leaves of the lower plants. The fallen parts of plants are consumed by insects and fungi, which are, in turn, eaten by some of the smaller animals on the ground. Whereas there are many large ground mammals feeding on plants in Malaya, they are rare in Queensland; two wallabies and a giant flightless bird, the cassowary, are the only ones. Probably as a consequence, no large carnivores have evolved in Queensland comparable to the tigers, leopards, and bears of Malaya. Other than the recently arrived man and his dog, the dingo, large snakes are probably the main carnivores in Queensland.

Thus the energy fixed by plants in the canopy is utilized by animals at all levels. Harrison points out that this is very similar to the sea, the canopy representing the phytoplankton in the lighted surface layers, with the animals feeding at all levels in both systems. Here we get the first inkling of another pattern. Different communities—even communities as dissimilar as those in a forest and those in the sea—have similar structures in the sense that they both have plants eaten by herbivores, which are in turn eaten by carnivores, and so on. These communities also have a geometrical arrangement with different plants, herbivores, and carnivores at the bottom and at the top. We shall see later that some of these similarities are quantitative as well as qualitative: the patterns of energy transfer and geometry are often nearly identical, even in communities whose species' names are utterly different. In other words, by suppressing the urge to describe communities by the names of their component species we find new patterns.

PATTERNS IN THE SEA

Predictions of the kind of life to be found in the sea are somewhat easier to make. The general circulation in the oceans carries heat away from the equator in the western parts of the oceans (along the eastern shores of the continents), so that tropical marine organisms are found up to about 30°N. and S. latitude. In contrast, the colder water from high latitudes flows towards the equator along the western shores of the continents, and tropical organisms, such as coral reefs, extend to only 10° or 15° latitude.

Because light is absorbed very quickly as one descends into the sea, plants grow only in the upper layer of approximately 80 meters (or less in turbid inshore waters). This means that attached plants grow only around coastlines, and floating plants only in the upper surface of the seas. All animals in the sea are dependent on the plant production in this thin upper layer. It is no wonder that the number of deep-sea animals is small in comparison to those at the surface. Because plants require certain mineral salts for growth, and because dead organisms tend to fall to the bottom, mineral nutrients must somehow be brought from the bottom into the lighted surface layers before plants can grow. This is no problem in shallow, turbulent coastal waters, but in the deep sea this mixing is a very slow process, and, as a consequence, plant productivity is low at the surface over deep oceans. The upwelling of water along steep coasts, which we described earlier, brings nutrients up and makes possible increased plant growth in such places as the ocean off California and Peru. We can draw an analogy between the vertical movements of water supplying nutrients in the sea and the vertical air currents supplying rainfall over land; both create areas of rich organic growth, although of course, for different reasons.

There is another effect of movement, which is probably more important in water than in air—the direct effect of turbulence. Whereas winds may directly affect plants and animals on land, the presence of tall vegetation effectively damps out all but the most destructive winds. However, because water is much denser than air, movement itself has a very great effect in the sea. This is especially true for animals in the plankton or on sandy beaches, where the few species present spend much of their energy keeping their positions in an unstable medium. The organisms living on rocks, however, have a stable substrate. If, in addition, these organisms grow sufficiently high above the rock surface so that there are spaces beneath them, a quiet zone is formed in which many species of more delicate animals live. Beds of mussels and seaweed reduce the water movement in the same way that forest trees reduce wind. Of course, some water movement is beneficial, for mineral nutrients and oxygen must be provided and wastes must be removed. Also, in very quiet water suspended particles settle out and may cover photosynthetic surfaces or clog respiratory organs.

These considerations enable us to construct the ideal physical environment for marine organisms: a firm substrate with a moderate amount of water movement in the upper lighted surface waters, with mineral nutrients in solution, and little seasonal variation in light, temperature, and salinity. We find this combination of characteristics in a tropical coral reef. Tidal and wind currents provide water movement, which is moderated by the complex structure of the living corals. The upward growth of corals keeps the reef in the lighted zone. The loss of mineral nutrients to the deep sea is reduced, because a coral atoll in deep water creates eddy currents, which tend to keep dissolved minerals in the vicinity of the atoll and to bring others up from the depths. Large green algae are attached to the dead coral skeletons, and smaller ones are contained within the cells and skeletons of the living corals. The metabolic wastes of the corals are not excreted into the sea, but are absorbed by these "internal" plants directly from the coral tissues. This cycling of mineral nutrients within the corals undoubtedly reduces their loss from the reef. As a result of all this integrated activity, the amount of energy passing through the plants and animals of a coral reef is the highest of any marine community, in contrast to the very low productivity at the surface of the deep tropical ocean nearby. Under these ideal conditions and because of the high productivity, the coral reef supports the most diverse and colorful assemblage of plants and animals to be found in the sea.

Of the diverse animals of the sea, three-fourths of the species live on firm substrates, such as rock or coral reefs, one-fifth on more extensive areas of sandy and muddy bottom, and the remaining one-twentieth in the plankton that occupies most of the oceanic space (Thorson, 1957).

PATTERNS IN FRESH WATER

The sea can be regarded as a permanent, continuous, stabilized environment; in contrast, streams and lakes are short-lived, isolated into small patches, and variable. Temperatures fluctuate severely in mid-latitude lakes. Also, as we described earlier, the water in deep lakes becomes "stratified" in summer, effectively sealing off the depths from all contact with air. Organic debris washed into the lake, and the animals and plants that die in the upper layers, sink to the bottom. Aerobic decomposition may completely use up the dissolved oxygen in the bottom water. Thus animals living on the bottom of a lake must tolerate extreme changes in oxygen supply between summer and winter, a stress not imposed on animals in other habitats.

Fresh water may also be variable in its dissolved chemicals. An organism being moved from one lake to another may pass from almost pure water to water more saline than that of the sea or to waters with completely different proportions of ions. For example, calcium content may vary greatly from lake to lake, and the fauna of "soft" water lakes is often very different from that of lakes with much calcium. There is also great variation in the size and degree of isolation of lakes. Small lakes, ponds, and streams, which are more likely to dry up than larger ones, present the organisms with a choice of hibernation or emigration. Thus fresh water seems to share the disadvantages of the land's variable climate and the sea's low supply of oxygen, as well as the isolation of islands.

Some freshwater organisms probably evolved from the invaders from the sea: algae, bacteria, fungi, and most of the invertebrate animals. However, one large group of animals—the insects—probably invaded from land (Macan, 1963). Their wax coatings, which keep moisture in on land, serve equally well to keep salts in or water out, and thus they do not expend much energy maintaining an osmotic balance, in contrast to the other organisms mentioned. Many insects return to the surface to breathe or to take a supply of air down with them. Because many insects can fly, they are able to emigrate if the water dries up. None of these adaptations is particularly advantageous in the sea, which possibly explains why so few insects have invaded it.

THE PATTERNS OF SINGLE-SPECIES POPULATIONS

Having discussed the patterns of whole communities, we shall now look more closely at the geography of single populations or species to see what determines their geographical boundaries.

The individual organism can live and reproduce only within a certain range of values of the physical factors in its environment. This has been called its "tolerance range." The tolerances of different individuals vary to some extent, so that the range of tolerance of a local population naturally is larger than any one of the individuals in it.

It is obvious that a population cannot exist under conditions that are outside its tolerance range. These conditions may be caused by purely physical occurrences, such as the climate, the topography, or the effect of geological "parent materials" on soils. However, even if these conditions are favorable to a particular population, they may be so altered by other kinds of organisms that they are no longer within the tolerance range of the population. If a population of small plants does not tolerate shade, it is excluded from an area shaded either by an overhanging rock or by a grove of trees.

Populations are not limited solely by their tolerance ranges. Direct effects of other kinds of organisms, such as herbivores, predators, and parasites, or direct interference by competitors may also set the boundaries.

Moreover, a population may not have actually stopped at its present boundary; it may still be spreading, its present limits being determined by a slow rate of dispersal. On a very broad geographic scale this is unlikely; the distribution of species with specialized dispersal mechanisms does not seem to be any wider than the distribution of those that lack them. *Littorina saxatilis,* an intertidal marine snail whose offspring, protected by the parents, do not spread far, has as wide a geographic range as other species of *Littorina* which release their young into the plankton, where they may drift over great distances. However, on a local scale, the rate of dispersal may determine boundaries, if there are frequent disturbances, such as floods or landslides, or if the climate is very changeable, as on the edges of a desert.

As an example of the ways in which a population may be limited, we consider the marine animals and plants of the rocky intertidal region. This is a particularly good place to study distribution, because the physical conditions change drastically in a short vertical distance from complete submersion to complete emersion. Also, the animals are quite sedentary, so that they can be moved around to see how they act under different conditions. Here we discuss only the upper and lower vertical limits, not the geographic horizontal ones.

Many studies have shown that species which extend higher on the shore are more tolerant of heat and desiccation than are those whose upper limits are below them. They are probably living to the limits of their tolerance range at the upper boundaries of their distribution. However, this is decidedly *not* the case for their lower limits. Most intertidal marine organisms can tolerate continuous submersion, if they are protected from predators and from species that may grow above them. In fact, many of them will grow faster and reproduce more quickly in continuous immersion.

Because they are not usually found below the intertidal zone but neverthless can tolerate its conditions, what determines their lower limits? In the few instances in which experiments have been performed the limits are clearly due to exclusion by competition with similar species in the "sublittoral zone" or elimination by its host of predators. Most of the larvae of intertidal species swim near the surface when they reach the stage of attachment, so that few of them ever attach below the intertidal zone. An even more extreme example of such specialized behavior was seen by Connell (1961*b*) in Scotland. The lower limit of the adults of the barnacle *Chthamalus stellatus* is near the very top of the intertidal range; in the greater part of the intertidal range below this they are eliminated in competition with another species of barnacle. The larvae do not attach below the middle of the intertidal zone, as shown in Fig. 1-24.

Can we apply these findings about animals on intertidal shores to the geographical distribution of all organisms? At the present time the answer is probably yes, but the amount of experimental data is small. Many species of plants are evidently living to the limits of their physical tolerance ranges in certain parts of their ranges. Tests, however, have mainly been made on the borders that extend into more rigorous climatic conditions. For example, the upper limit of a species on a mountain is probably set by its tolerance to the rigorousness of the winter weather. The low herbaceous and shrubby alpine tundra is also prevented from extending to lower elevations, but not by an inability to withstand the somewhat warmer conditions 500 feet lower. It simply cannot grow in the dense shade under the trees of these lower slopes. In other instances it may not be able to compete with faster-growing grasses. Its upper limit is set by its tolerance range

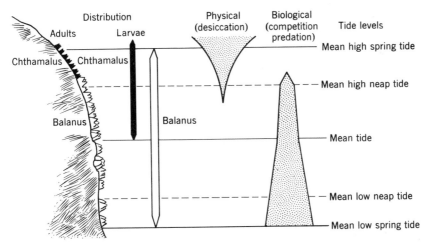

Fig. 1-24. Factors that control the distribution of two species of barnacles in an intertidal gradient. The young of each species settle over a wide range but survive to adulthood only within a more restricted range. Physical factors such as desiccation control upward limits of *Balanus*, whereas biological factors such as competition and predation control downward distribution of *Chthamalus* in the lower portion of the intertidal gradient where physical environment is less limiting. This model can be considered to apply to more extensive gradients such as an arctic-to-tropic or a high-to-low altitude gradient. [Redrawn by Odum (1963) from J. H. Connell (1961b).]

to purely climatic conditions, and its lower limit by competition with other species, which either alter the physical conditions or interfere directly. The necessity to live at these high elevations has naturally led to a selection of certain traits which are so specialized that the species may not be able to reproduce if it is transferred to conditions that appear to be more favorable. Thus the temperate alpine tundra plant *Oxyria digyna,* does not begin its spring growth until the length of daylight reaches about 15 hours (Mooney and Billings, 1961). This control by "photoperiod" rather than by temperature means that the plant does not begin to grow during an unseasonable warm spell in early spring, which may be followed by a killing frost. Only when summer has arrived, and the days have lengthened to 15 hours, is the danger of later frosts reduced. If this plant were transferred directly to an equatorial mountain where the days never reach 15 hours, it would not grow even if the climate were more favorable; the investigators duplicated the equatorial climate conditions in a greenhouse and found that, as predicted, the plant did not grow.

Up to this point we have dealt with the broad geographic limits of a population. However, within a population there is also a pattern of distribution of the individuals, which is seldom random. Usually the individuals are closer together than we would expect in a random pattern, being aggregated into clumps, but sometimes they are further apart in a rather uniform distribution.

Individuals may be clumped in favorable places when they are excluded from the rest of the region by the factors we have just discussed. On the other hand, they may occur in social groups such as herds or schools, so that at any one time they are inhabiting only a part of a uniformly favorable area.

Uniform distributions, sometimes shown by the trees of dense forests, breeding birds, and certain other organisms (Fig. 1-25), must be the result of some kind of negative interactions between the individuals. Competition between the parts of a cell or an organism can also result in a uniform distribution; the hairs on your hand are uniformly spaced.

Thus the pattern distribution of individuals within a population can provide clues to their behavior; this may be useful in studying populations, such as the animals on the bottom of the deep sea where observations are difficult to make.

A Footnote on Methodology. Here, and elsewhere in this book, we talk of "clumped" organisms or of "uniformly distributed" or of "randomly dispersed" organisms. One of the interesting features of human experience is that, shown a truly random distribution of points (such as are made approximately by falling raindrops), people think that it is "clumped" (Fig.

Fig. 1-25. Nearly uniform spacing of creosote bushes (*Larrea*) in an Arizona desert. (Photograph by M. Mathias, University of California, Los Angeles.)

1-26 and 1-27). Or, if asked to put points on a paper in a "random" pattern, they usually space the points too uniformly. (Of course, it is well-nigh impossible to construct a *truly* random pattern deliberately.) How does an ecologist recognize a random pattern when he sees one? The answer is this also explains how he recognizes a clumped or a more uniform-than-random pattern, for departures from randomness extend in the direction of patterns either too uniform (orchard-like) or too clumped. To recognize a random

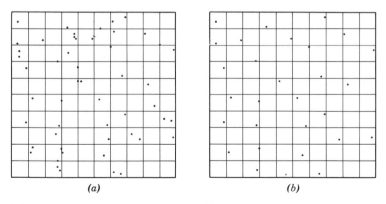

(a)   (b)

Fig. 1-26. (a) Fifty random points from table of random numbers. (b) Orchard-like pattern formed by eliminating crowded points in (a) (competition between nearby points).

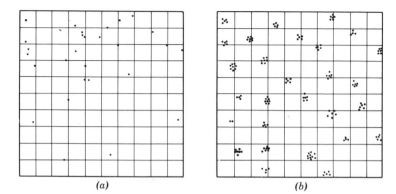

Fig. 1-27. (a) One pattern of clumping formed by removing most of lower points from Fig. 1-26a (poor soil in lower part). (b) Another pattern of clumping formed by letting points in Fig. 1-26b give rise to many new ones (distribution of seeds).

pattern in space, the usual procedure is to subdivide the space into a checkerboard of unit squares and to count and record the number of squares that are empty, have one point, two points, three points, and so on. We also count the total number of points and the total number of squares from which, by dividing, we can calculate the average number of points per square. It turns out that this is the only number we need to construct the expected proportion of squares that are empty, have one point, and so on. The reasoning is as follows:

1. The density of points is m per square; thus the probability that a tiny area, dh, has a point is $m\,dh$ (the chances of such a tiny area having more than one point are negligible)

2. The probability that the tiny area dh is empty then is $1 - m\,dh$.

3. Now we assume the probability that a larger area, h, is empty and, because we wish to let h vary, we give the probability a symbol, $P_0\,(h)$, to be read as the probability of zero in area h. We see that the probability of zero in area $h + dh$ is the probability that there are zero points in h and also zero points in the additional area $dh;$ that is, it is the product of the probability of zero in h times the probability of zero in the additional dh. Symbolically, this is written as

$$P_0(h + dh) = P_0(h) \cdot P_0(dh)$$

However, we have already said that $P_0(dh) = 1 - m\,dh$. By substituting this value, we obtain

$$P_0(h + dh) = P_0(h)(1 - m\,dh) = P_0(h) - mP_0(h)\,dh \qquad (1)$$

This can be written

$$\frac{P_0(h + dh) - P_0(h)}{dh} = -m\, P_0(h)$$

and, when dh is small, the left side is approximately the derivative, $dP_0(h)/dh$, of $P_0(h)$

$$\frac{dP_0(h)}{dh} = -m\, P_0(h)$$

This can be written

$$\frac{1}{P_0(h)}\frac{dP_0(h)}{dh} = \frac{d\log_e P_0(h)}{dh} = -m$$

so that, by integrating, we obtain

$$P_0(h) = e^{-mh} \qquad (2)$$

(The constant of integration is zero, as we see by setting $h = 0$ and remembering that $P_0(0)$ must be 1.) Finally, to find the probability that one of our *unit* squares is empty, which is the same as finding the expected *proportion* of squares that are empty, we set area h equal to 1 and obtain $P_0 = e^{-m}$. This is the first part of our answer: a fraction e^{-m} of the squares is expected to be empty if the distribution is random. The reader may ask where randomness enters the derivation; the answer is that random means that each small area dh has an equal probability of containing a point, irrespective of whether adjacent small areas do or do not contain points. This is the property we used in deriving Equation 1.

4. Now we can easily extend our results to find the probability that a square has one, two, three, or more points. Let us first take $P_1(h + dh)$—the probability that an area $h + dh$ has one point. This can happen either if h has the point and dh is empty or if h is empty and dh has the point. In symbols, using previous results, we have

$$\begin{aligned} P_1(h + dh) &= P_0(h)\, P_1(dh) + P_1(h)\, P_0(dh) \\ &= P_0(h)m\, dh + P_1(h)\,(1 - mdh) \\ &= me^{-mh}\, dh + P_1(h) - m\, P_1(h)\, dh \end{aligned}$$

which can be written

$$\frac{P_1(h + dh) - P_1(h)}{dh} = me^{-mh} - mP_1(h)$$

When we write the left side as a derivative, as done previously, we obtain

$$\frac{dP_1(h)}{dh} = me^{-mh} - m\, P_1(h)$$

and

$$P_1(h) = mhe^{-mh}$$

can be verified by substitution to be the solution of this. In other words, the proportion of unit squares, which are expected to have one point, is me^{-m} (setting $h = 1$). Next we turn to the probability that an area $h + dh$ has *two* points. By the same procedure, this satisfies the equation

$$\frac{dP_2(h)}{dh} = mP_1(h) - mP_2(h)$$

whose solution is

$$P_2(h) = \frac{(mh)^2}{2} e^{-mh}$$

so that the expected proportion of unit squares with two is $(m^2/2)e^{-m}$. Proceeding in this fashion, we find the following rule.

RULE. *Randomly scattered points with an average density of m points per unit area will be distributed in unit areas as follows:*

Squares containing	0	1	2	3	. . .	n points
Expected proportion of such squares	e^{-m}	me^{-m}	$\dfrac{m^2}{2}e^{-m}$	$\dfrac{m^3}{6}e^{-m}$	. . .	$\dfrac{m^n}{n!}e^{-m}$

This is known as the Poisson distribution. It tells us what we *expect* if the distribution is truly random, but not what we would necessarily *always* find. (We expect a coin to turn "heads" half the time, but it will not necessarily do so.) How can we tell whether a departure from the expected proportions is sufficiently small to be consistent with randomness or whether it is so large that we prefer to consider it basically nonrandom? Statistics serve to answer such questions, and statisticians often use what is called the x^2-test for this purpose (x is the Greek letter chi); x^2 is defined in terms of the observed numbers O_i of squares containing each number i of points and the expected numbers E_i of squares containing each number of points

$$x^2 = \sum_i \frac{(O_i - E_i)^2}{E_i}$$

$$= \frac{(observed\ number\ of\ squares\ with\ 0\ points -\ expected\ number\ with\ 0\ points)^2}{expected\ number\ with\ 0\ points}$$

$$+ \frac{(observed\ number\ with\ 1\ point -\ expected\ number\ with\ one\ point)^2}{expected\ number\ with\ 1\ point}$$

$$+ \cdot\ \cdot\ \cdot$$

See Table 1-1 for an example. Notice that we lump the categories included in braces to make the expected number in each category add to at least five. (This is done for purely technical reasons and is essentially a limitation on the x^2-test itself—all expected proportions should be at least five.) Thus x^2 is a measure of departure of the observations from the "expected" values. One other number, beyond x^2 itself, is necessary for our purpose: the "number of degrees of freedom," K. This number K is simply two terms fewer than the number of terms in the sum defining x^2. (See Table 1-1 for clarification.) Normally, every term in a sum can vary independently, but our expected terms are matched to the observed ones in two ways: the totals match and the values of m match. Hence we lose two "degrees of freedom" from the number of terms in the sum. Statisticians have tables of x^2 which tell how often a departure as large as (or larger than) the observed departure would come about by chance alone. If it happened rarely, by chance alone, the scientist would usually conclude that the assumption of randomness is wrong. To avoid consulting statistical tables, we can keep in mind the crude rule that if x^2 (with K degrees of freedom) is greater than about $K + \sqrt{2K}$, we are justified in suspecting that the distribution is not random. From Table 1-1, for example, we see that, because 4.39 is much less than $8 + \sqrt{16} = 12$,

Table 1-1   Number of Yeast Cells in Squares of a Hemocytometer
(From data of "Student")

Number of Yeast Cells in Square, i	Observed Number of Squares, O_i	Expected Number of Squares, E_i
0	0 } 20	3.7 } 21.1
1	20	17.4
2	43	40.6
3	53	63.4
4	86	74.2
5	70	69.4
6	54	54.2
7	37	36.2
8	18	21.2
9	10	11.0
10	5 } 9	5.2 } 8.7
11	2	2.2
12	2	0.9
>12	0	0.4
	400	

Total number of yeast cells is $\Sigma i O_i = (0 \times 0) + (1 \times 20) \times (2 \times 43) \cdots + (12 \times 2) = 1872$

$$m = \frac{1872}{400} = 4.68$$

$$e^{-m} = e^{-4.68} = 0.009276$$

$$x^2 = \frac{(20.0 - 21.1)^2}{21.1} + \frac{(43.0 - 40.6)^2}{40.6} + \cdots + \frac{(9.0 - 8.7)^2}{8.7} = 4.39$$

10 terms in sum: $10 - 2 = 8$ degrees of freedom

there is no reason to assume that yeast cells are not distributed randomly. However, if x^2 is larger than $K + \sqrt{2K}$, how can we tell whether the departure from randomness is in the direction of clumping or whether it is orchard-like? Usually it is sufficient to compare the empty squares. If the observed number of both empty squares and squares with many points is smaller than the expected number, it reflects orchard-like distribution; if the observed proportion is greater than expected, it reflects clumping. An orchard-like distribution is very revealing and usually reflects a competitive repulsion; a clumped distribution is more complicated and may reflect a patchy environment, in which the organisms prefer one kind of patch, or it may reflect gregarious organisms. (See Figs. 1-27, 1-26).

If the distribution is clumped, m, the mean density is misleading as a measure of how crowded the animals are. That is, if the animals are constantly moving around, they encounter one another more frequently in a clumped distribution than they would in a random distribution having the same mean density. For the case in which the positions of the individuals are random within the clumps, Lloyd (1966) has suggested another measure, $\overset{*}{m}$, the "mean crowding," defined as the mean number *per individual* of other individuals currently sharing the same unit square. The relationship between "mean crowding" and mean density can be seen as follows: consider a total population of N individuals having a clumped distribution in Q unit squares. For identification, number each individual with the subscript i ($i = 1, 2, 3, \ldots, N$) and each unit square with the subscript j ($j = 1, 2, 3, \ldots, Q$). Let X_i be the number of other individuals in the same unit square, determined for each individual separately.

Thus, for an individual by itself, $X_i = O$; two individuals in the same unit square contribute two X_i values of 1; 3 individuals together contribute 3 values of 2, and so on. The "mean crowding," or average value of the X_i's, will therefore be

$$\overset{*}{m} = \frac{\sum\limits_{i=1}^{N} X_i}{N} = \frac{\sum\limits_{j=1}^{Q} x_j (x_j - 1)}{N}$$

where the x_j's are the numbers in each of the unit squares and $\sum x_j = N = mQ$. Now we have

$$\overset{*}{m} = \frac{\sum x_j^2}{N} - \frac{\sum x_j}{N}$$

$$= \frac{(1/Q) \sum x_j^2}{m} - 1 + m - m^2/m$$

$$= m + \frac{(1/Q) \sum x_j^2 - m^2}{m} - 1$$

$$= m + \frac{(1/Q) \sum x_j^2 - (2m/Q) \sum x_j + Qm^2/Q}{m} - 1$$

and, because $Qm^2 = \sum_{j=1}^{Q} m^2,$

$$\overset{*}{m} = m + \frac{(1/Q) \sum (x_j^2 - 2mx_j + m^2)}{m} - 1$$

$$= m + \frac{(1/Q) \sum (x_j - m)^2}{m} - 1$$

Students who have studied statistics will recognize the numerator of the second term as the mean-squared deviation from the mean, called the "variance," usually symbolized σ^2. Thus, we have

$$\overset{*}{m} = m + (\frac{\sigma^2}{m} - 1)$$

In a truly random distribution, the mean and variance of the x_j's tend to be equal (we omit the proof here, but it can be found in many statistics books), so the quantity in parentheses tends to disappear, and $\overset{*}{m}$ is identical with m, as one would expect. On the other hand, it is a general property of clumped distributions that the variance exceeds the mean. (This should be intuitively obvious to the student in view of the great variability from place to place in the clumped distributions in Figs. 1-26 and 1-27). Thus, for example, in a situation where $m = 10$ and $\sigma^2 = 100$, we have $\overset{*}{m} = 10 + ((100/10) - 1) = 19$. In this distribution, the animals encounter $\overset{*}{m}/m = 19/10 = 1.9$ times as many other individuals around them, on the average, as they would if the distribution were random. Lloyd proposes this ratio, $\overset{*}{m}/m$, as a measure of "patchiness" and gives methods for dealing statistically with sample estimates.

OTHER PATTERNS IN SPACE

Another spatial pattern is the number of species on islands. Remote or small islands have fewer kinds of plants and animals than larger ones or those nearer sources of colonization. We might suppose that, given enough time, the remote or small islands would eventually become similar to the others. It is not certain, however, that history is always the explanation. To understand this, let us consider a remarkable event.

On August 7, 1883, the island of Krakatau in the strait between Java and Sumatra exploded. It was no ordinary explosion: the whole island was blown to bits, part of it disappearing completely (or, more correctly, being turned into fine dust, which caused red sunsets around the earth for months) and the remaining part being covered with hot volcanic debris to a depth of two hundred feet. Two months later it was still hot enough to turn rain into steam; quite likely all life on the island perished; perhaps some underground worms or other soil forms survived. In any case, conspicuous higher organisms were completely eliminated, and a beautiful experiment, demonstrating the powers of plants and animals to recolonize across open salt water, was begun. Most of the recolonizers came from Java, twenty-five miles away, and Sumatra, fifteen miles away. One year after the explosion only a few blades of grass and one spider were found (Dammerman, 1948). When the first major expedition visited Krakatau in 1908, 202 species of animals (exclusive of microscopic forms) were discovered in a three-day search. A sixteen-day search in 1919 produced 621 species, and fourteen days in 1934 (fifty years after the eruption) produced 880 species. By this time a young forest covered much of the island, and thirty species of birds had moved in, most of them eating some of the more than 700 species of insects that abounded. Four species of mammals had arrived, and six species of reptiles, including the python, were found. Krakatau, by 1933, had about one quarter of the number of species of terrestrial vertebrates found on nearby islands of comparable size, but nearly two-thirds of the expected number of insects. In contrast to other vertebrates, birds had virtually reached the expected number of species even before 1933. Clearly, the birds and insects were more rapid colonizers than the mammals. Other invertebrates (e.g. spiders and mollusca) had 92 per cent as many species on Krakatau as on comparable undamaged islands. Dammerman concluded that the Krakatau fauna would be complete "not after centuries, but within a comparatively short time, much sooner than we ourselves had expected."

How are we to interpret this case history of Krakatau? Fundamentally, it seems to show that plants and animals can cross open stretches of ocean remarkably easily. Even organisms not supposed to live in salt water seem capable of this feat. It provides good evidence of what Simpson (1965) calls "sweepstakes dispersal," in which each species has a positive, although small, chance of crossing in any given year. The most remote islands, such as the Hawaiian group, New Zealand, the Galapagos, and Tristan da Cunha, seem to have been exclusively colonized in this way. It certainly is possible that they were, as Krakatau shows; no other feasible way has been suggested. (People have often postulated a land bridge connecting each island to the mainland. Although islands near the mainland were often so connected, there is no reason at all to believe that the remote islands were.)

Larger continents, however, have been more closely connected. The Isthmus of Panama, Arabia, and the chain of islands (or even isthmus when sea levels were lower) that bridges the Bering Strait, connect the major continents and have allowed, at one time or another, fairly free interchange of living forms. This has permitted the major continents to acquire rather complete and well-balanced faunas and floras, in contrast to the remote islands, which either have striking gaps or startling examples of makeshift. For example, woodpeckers of considerable variety have become expert at finding insects under bark and

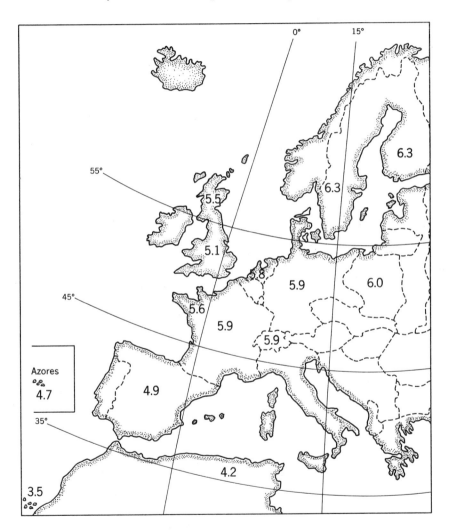

Fig. 1-28. Average clutch of the Robin (*Erithacus rubecula*) in different countries (from Lack, 1954).

even in the wood of trees; they are found on the major continents. The remote Galapagos islands, colonized by chance, never acquired woodpeckers, but evolved a surprising substitute in one of Darwin's finches, a group that did get to the Galapagos and then speciated to such an extent that one type or another took up nearly every way of life found on the mainland (Lack, 1947). The woodpecker finch, not having evolved the complicated morphology of true woodpeckers, has learned to pick up a cactus spine and use it to prod into crevices in the way of a woodpecker's pointed bill. When the finch dislodges an insect, it drops its cactus spine, picks up the insect, and eats it.

Turning to another topic, most animal species vary in size, shape, and color from one edge of their range to another. In general (but by no means always), the forms at the northern edge of a range are larger and chunkier and those in wet areas are darker. Why should this be?

Not only are there differences in appearance among individuals from different places, but there are also differences in physiology. Tropical forms, for instance, usually have a reduced birthrate. Figure 1-28 shows the number of eggs usually laid in different regions by the European robin (*Erithacus rubecula*). Why this change in birthrate?

In fact, as we think about the phenomena that show patterns in space, it becomes clear that virtually all phenomena do! Certainly any process

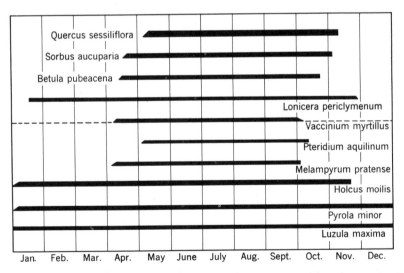

Fig. 1-29. Seasons of active assimilation of plants in a British oak wood. The species are ranked from tallest at the top to lowest herbaceous cover at the bottom. (After Salisbury, 1925.)

controlled by events that vary from place to place does itself show a spatial pattern. This is the virtue of looking at patterns in space, for the very shape of the spatial pattern often suggests the environmental factors that are in control. The same fact—that most phenomena differ from place to place, hence show a spatial pattern—makes it impossible for us to review all such patterns. Therefore, we stop here and turn to a sample of patterns which have more of a time component.

We begin with a striking example due to E. J. Salisbury (1925), who published several accounts of the structure of British woodlands. Salisbury noticed that as the spring progressed green foliage and flowers appeared higher and higher in the woodland (Fig. 1-29). Herbaceous vegetation appeared mostly in leaf and carried out its reproduction early in the spring while frosts were still frequent and the sun low in the sky. Bushes appeared in flower a little later, when conditions were better for carrying out photosynthesis and more insects were about for pollination, but bushes, too, completed their flowering before temperature and light had reached "optimal" levels. The trees alone apparently were able to choose their seasons freely and elected to flower when the sun was high and the danger of frost had passed.

The last pattern to be discussed is one shown by the fossil history and the present distribution of organisms. From these we can infer (with a good deal of uncertainty) how the pattern gradually developed during past time. For example, all trees of the beech family occur in the northern hemisphere, except one genus, *Nothofagus*, the Antarctic beech. Figure 1-30 shows where these trees occur now and where fossil pollen has been found; both are widely distributed in the southern hemisphere, and fossil pollen has also been found in Asia.

Because most of the beech family occurs in the northern hemisphere, and fossil pollen of *Nothofagus* also occurs there, the biogeographer Darlington (1965) has inferred that *Nothofagus* arose from an ancestor in Asia and spread to the southern hemisphere, probably first to Australia. Once there, seeds could have blown or drifted in the strong westerly currents to New Zealand and South America. Antarctica could have served as a "stepping stone" to South America; fossils from the same period as *Nothofagus* pollen, including plants that are usually found in cool temperate forests, have been found in the shallow sea at the edge of the Antarctic peninsula. This is fairly clear evidence that such a forest once existed there; thus it is possible that the rest of the border of the continent may once have been more hospitable to *Nothofagus* trees than it is now. We have given this example to show how difficult it is to reconstruct the past history of populations. Most other examples present even less evidence!

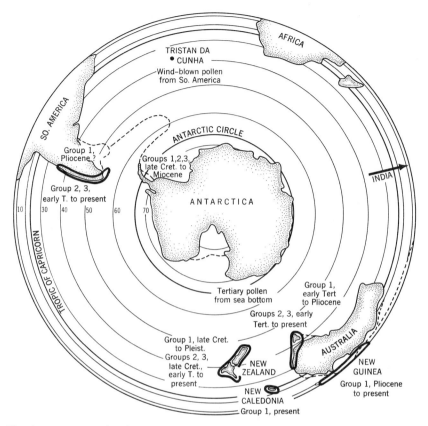

Fig. 1-30. Present distribution of southern beeches of the genus *Nothofagus* and summary of the pollen record of the three groups of the genus. (After Darlington, 1965.)

PATTERNS "ELSEWHERE"

The title of this chapter is "Patterns in Time, Space, and Elsewhere." What patterns are there that do not appear in time and space? In one sense, all patterns occur in time and space, because time and space are the dimensions of our world. Often, however, the most interesting patterns become obvious only as abstractions, in which time and space are incidental. We have already considered some of these patterns. For example, we noted that all communities have plants eaten by herbivores, herbivores eaten by carnivores, and so on. If we measure the number of new calories produced and stored in the bodies of plants, herbivores, carnivores, etc., per year in a square meter, we can plot them in a "pyramid of productivity," as shown in Fig. 1-31. This is an abstract pattern in the sense that its components are mental constructions obtained by adding various numbers. As another simple-minded example, we note that birthrates change with latitude (Fig. 1-28). In the case of the robin this is directly observable as the number of eggs in a nest. In general, however, the pattern is more obscure, because "birthrate" is not itself an observable number, but rather a number derived after some prescribed calculations. It turns out that the most interesting patterns in population biology are abstract ones. This is, in fact, the way science works: we find simple relations (patterns) among the objects that we actually see; then we find more basic relationships among the patterns that we discovered first; then even these higher relationships appear to be connected by even more basic relationships, and so on. The more sophisticated and interesting a science becomes, the more the patterns appear to relate entities that we do not observe directly.

In this introductory chapter we have discussed some of the obvious patterns that appear when we view the distribution of actual individual organisms in time and space. These patterns need to be understood, of course, but, as we proceed, we shall find more and more patterns of the more obscure variety, which relate these concepts.

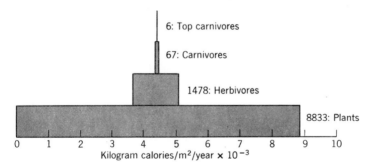

Fig. 1-31. The "pyramid" of productivity in a community (Silver Springs, Florida). (After H. T. Odum, 1957.)

PART TWO

THE EVOLUTION
OF POPULATIONS

2

The Process of Natural Selection

HISTORICAL INTRODUCTION

On November 24, 1859, the world apparently was ready for Darwin's *On the Origin of Species;* although John Murray was the only willing publisher Darwin had found, the first printing sold out in one day. The idea of evolution by natural selection did not, however, come to Darwin full-fledged in 1859. In fact, the *idea* of evolution had had an honorable history since the days of the Greek atomists. But Darwin made evolution seem so likely that he shifted the burden of proof to those who chose not to believe it.

How did Darwin come to believe in evolution? Was he a born radical, who always took the unorthodox view? Nothing could be farther from the nature of his boyhood. As a boy he seemed unusual only in the enthusiasm with which he collected beetles and shot birds. He went to Edinburgh to study medicine, but was so revolted by the dullness of the lectures and the sight of an operation that he left. He then went to Cambridge, planning to enter the ministry. He later* said: "I did not then in the least doubt the strict and literal truth of every word in the Bible." At Cambridge he made a mediocre record but he did impress some of the more perceptive dons, who eventually changed his life by recommending him for the post of naturalist on the voyage of HMS *Beagle*, which was under orders

---

* Darwin's autobiography (see bibliography), written for the enjoyment of his grandchildren, makes very charming reading and gives a candid account of his life.

especially to survey parts of South America. The five years he spent on the *Beagle* (December 2, 1831 to October 29, 1836) transformed him into a superb scientist and provided a rich background of experience, on which he would draw during the remainder of his life.

He made lasting contributions to every subject he studied thereafter—from the formation of coral reefs to the fertilization of orchids and from the action of climbing plants to the study of animal behavior. *On the Origin of Species,* however, remained his greatest work.

Darwin had been advised to take Lyell's *Principles of Geology* with him on the voyage, but by no means to believe it all. It espoused the view that the physiography of the earth could be understood in terms of the action of present-day phenomena. No cataclysms of an unknown or unique kind were required to produce the mountains, oceans, lakes, and smaller structures that we see about us. Rather, in modern terms, Lyell's book showed how the present-day earth evolved from an earlier uniform one under the action of known agents, such as glaciers, gradual rise in land level, and winds. Darwin traveled with Lyell's book for an intellectual companion, with great energy, boundless interest, and with an unequaled ability to observe and describe objectively what he saw. He gradually became convinced that evolution gave a simpler explanation of the patterns of life than did any alternative. He spent the next twenty years gathering more data and planned a detailed account of evolution. When Alfred Russell Wallace came to the same conclusions independently, Darwin's friends persuaded him to publish, simultaneously with Wallace, their common views on evolution by natural selection (Darwin and Wallace, 1958 edition). This 1858 essay received little attention, and in 1859 Darwin published an abstract of his big work. This abstract is *On the Origin of Species,* and the "big" work never appeared.

Although Darwin was ignorant of modern genetics, his account of natural selection is still very fresh and is still up to date in its essence. In particular, the chapters on "Struggle for Existence" and "Natural Selection" are the cornerstones of the material in this book.

MECHANISMS OF NATURAL SELECTION

There is a singularly close analogy between natural selection and bank or stock investments. Imagine an investor who deposits some silver money and some paper money in a bank which pays 3 per cent interest on paper money and 4 per cent interest on silver money. If the interest is paid in the same currency as the investment (as stocks often are), so that there is "heredity," that is, silver begets silver and paper begets paper, the quantity of silver money in the bank account increases faster than the amount of paper (silver is "fitter" than paper); therefore, the proportion of money that is silver increases, and the proportion that is paper decreases. This is an almost exact counterpart of natural selection in an expanding population of haploid* individuals, in which like effectively begets like. In this case, silver gains its advantage by differential "reproduction." Alternatively, consider the case in which both silver and paper earn 5 per cent interest, but the investor spends 2 per cent of the principal of paper money each year after the interest is paid and only 1 per cent of the silver. Then, exactly as before, the paper money in the bank account will increase at 3 per cent and the silver at 4 per cent (which is therefore "fitter"), and the proportion that is silver will gradually grow toward 100 per cent. This is analogous to selection by differential "mortality" in an expanding population.

If the investor spends an amount equal to the total interest paid, his account remains constant in size. However, unless he spends paper and silver at exactly the rates at which their separate interests accumulate, either silver or paper will increase at the expense of the other. This is the analog of selection in a constant-sized population of haploid individuals. (We shall return to this later in a quantitative form.)

Again, we can consider a hypothetical investor, who prefers silver money and who wants to maintain as much of his investment in silver as possible. However, his bank always pays some of his interest in paper and other nonsilver currency. The investor is careful to spend the paper and other nonsilver currency in preference to silver. By so doing, he keeps his account predominantly silver, in opposition to his bank's efforts to broaden his investments. This is analogous to the "stabilizing selection" by which well-adapted species preserve their genetic virtues by weeding out extremes.

Suppose now that the heredity is less perfect and that not all of the interest is paid in the currency of the investment. A little reflection shows that the degree of "heredity" in paying the interest determines whether one currency becomes predominant. If there is no "heredity" and the interests on both the paper and silver investments are paid in the same proportion of silver and paper (e.g., if all interest is paid half in paper and half in silver, irrespective of the type of investment),

---

* For genetic terminology, see Stern, H., and D. Nanney, *The Biology of Cells,* Wiley.

neither paper nor silver can progressively take over. They will, however, reach an equilibrium (50:50 in our example), and the different interest *rates* will have no effect on this final proportion. Thus selection cannot act in the absence of heredity, but if there is a strong correlation between the currency of the investment and that of the interest, the selection continues, but not in the simple fashion outlined. In populations of real diploid individuals the offspring are not genetically identical to the parents—for example, heterozygous parents have some homozygous offspring—but there is often a strong correlation between parents and offspring. Hence our qualitative picture is still fine, but we must expect some complications. These complications are discussed in a later section on the quantitative genetics of populations, but first we must complete the qualitative picture.

We call the interest rate minus the rate of spending the "fitness." It was 3 per cent for paper and 4 per cent for silver in the first two examples (in expanding bank accounts). In the case of the constant total bank account the fitnesses are not zero, however, for, if the silver assumes control, the paper decreases at a rate just equal to the increase of the silver, so that the total is constant. Hence the paper currency has a negative fitness and the silver a positive fitness. Only the average fitness of the whole investment is zero. In real populations the "fitness" is the birthrate minus the death rate. (All births and all deaths are counted, even if the offspring are not genetically identical to the parents, although then we have to make allowance for the genetic constitution of the children).

Suppose that each normal robin in a population newly colonizing an island has a two-thirds chance of surviving and reproducing. Further, suppose each pair lays four eggs. Finally, suppose each bird, newborn or old, has the same two-thirds chance of surviving and reproducing the next year, and so on. The fitness of these birds is easy to calculate, for two-thirds of the previous year's birds reproduce and because each pair produces four young each individual parent can be attributed two young. Thus, for each bird present this year, there are $\frac{2}{3} \times 2 = \frac{4}{3}$ birds next year. We shall call this fraction of birds the "selective value" of the normal birds. Or, considered in another way, they are increasing at a $33\frac{1}{3}$ per cent interest rate. We shall reserve the term "fitness" for this interest-rate description of the population increase. We may find a few white robins in the population and inquire how their fitness compares with that of the normal birds. On studying the lives of a few of the white robins, we may suppose that more of them die before reproducing and that fewer mate, so that only one-third of the white birds alive one year reproduce the next.

However, we may find that those white birds that did reproduce (mating either with normal birds or with other white birds) reared the normal four-egg clutch, two of which can be assigned to the white bird (who has, after all, except for the sex chromosomes, contributed one-half to the genetic endowment of these four offspring). Each robin then leaves $\frac{1}{3} \times 2 = \frac{2}{3}$ descendant for the next generation, compared with a $\frac{1}{3}$ left by normal birds. We can also say these white robins contributed a negative $33\frac{1}{3}$ per cent interest rate. Finally, we can take the ratio

$$\frac{\text{selective value of white}}{\text{selective value of normal}} = \frac{\text{number of descendants per white bird}}{\text{number of descendants per normal bird}}$$
$$= \frac{2}{3} \div \frac{1}{3} = \frac{1}{2}$$

as the *relative selective values of white to normal*. (We are careful not to use relative per cent increase (fitness), which would give us a different number $-33\frac{1}{3}/33\frac{1}{3} = -1$.) Knowledge of the relative selective value turns out in most (but not all!) cases to be as useful as the full knowledge of the actual numbers of descendants or fitness. Now, after this digression on measuring fitness, we return to the theme of relating natural selection in nature to the growth of investments.

Evolutionists have used two slightly different ways of presenting the argument of natural selection. Darwin let the phenotypes of organisms be the investments that grow at different rates. However, phenotypes of organisms have imperfect heredity, partly because of the nonhereditary effects on the organism's phenotype and partly because diploid phenotypes do not uniquely determine their genotypes, hence do not determine their heredity. On the other hand it *is* the phenotype that determines the fitness. Thus Darwin described the progressive change of *phenotype* through the action of natural selection, but had to be vague about the hereditary nature of these phenotypes, hence about the actual progress of evolution. Modern geneticists have taken another view of natural selection. They have let the genes themselves be the investments, whose proportions are altered by their relative interest rates. The genes duplicate themselves perfectly in meiosis (except for rare mutations which we shall study in a later section) and have a nearly perfect heredity, which is in marked contrast to that of the phenotypes in Darwin's argument. However, we seldom get something for nothing, and the modern geneticists' argument substitutes precision of heredity for precision of control of fitness, for, although phenotypes control fitness exactly, the individual genes do not determine a precise rate of their own increase. Instead, a packaging problem develops, and the number of descendants left by a particular gene depends on whether that gene, in diploid form, is combined with another allele like itself or with a different allele. Even then, we can assign a fitness to those genes only if we know how the diploid genotype and its interacting genes control the phenotype—knowledge that we seldom, in fact, possess.

If we can relate phenotype to genotype, the two arguments become identical, for then we can assign a fitness to a genotype, by way of its phenotype, thereby providing a remedy for the geneticist's difficulty and also a heredity to a phenotype, thus curing the vagueness in Darwin's argument. However, we are far from possessing this knowledge, so that at present the arguments are still different, at least in practice, if not in theory. We shall proceed as far as we can with Darwin's argument first, and then, in a section on the quantitative genetics of populations, we shall pretend that we can assign fitnesses to genotypes and to the genes that compose them.

Any two individuals in a population differ from each other in various measurable ways. One may be taller, heavier, or able to run faster, or be quicker at solving some problems, have differently colored eyes, produce calluses with less abrasion, have a different sex, be more fertile, or differ from the other individual in any one of an endless variety of measurable ways. Similarly, two plants may differ in the size of seed from which they grew, earliness of flowering, number or color of flowers, height, and so on. The set of all (at least theoretically) measurable characteristics of an individual, throughout its lifetime, is called its *phenotype*. (We normally exclude characteristics that can be measured only by breeding experiments, which we call genotype by contrast.) Clearly, a phenotype cannot be specified completely, because we cannot enumerate all of the measurable characteristics. However, it is often practical to enumerate differences among individuals. Two corn plants may differ in time of flowering, and for convenience we are likely to refer to the early flowering phenotype and the late flowering one without mentioning all of their other characteristics. Thus we can describe differences in phenotype between two individuals and, as seen in the last paragraph, we can compare their success at leaving descendants—their fitnesses. It is a general rule that *most different phenotypes have different fitnesses*. This is the most important empirical generalization behind natural selection and is therefore worth considering more fully. It is easy to illustrate. Consider, for example babies born in London hospitals. Karn and Penrose (1951) have shown that among 13,730 babies, the light phenotypes (weighing less than, say, five pounds) and the heavy ones (more than ten pounds) were more likely to die before reproducing than were the medium-sized ones of about seven or eight pounds. Thus the fitness of the medium-sized babies was greater than that of either the heavy or the light ones, unless the medium ones have a compensating reduction of birth rate. To demonstrate the truth of the italicized statement, it is only necessary to examine a large number of pairs of differing phenotypes and to show that in most cases their fitnesses do differ. Among the parts thus examined there is no clear-cut case of differing phenotypes without different fitnesses, so that it is probably permissible to say that *all* different phenotypes have different fitnesses. For example, it used to be argued that different blood types—A, B, AB, and O—must have no effect on fitness. Recently, however, it has been found that type O is more susceptible to ulcers than are the others. Other types are associated with additional diseases. Hence even these characters which have no visible effect must alter the fitness. However, the argument for natural selection requires only that there exist phenotypes of differing fitnesses, so that even if the statement is someday disproved by finding two phenotypes without detectable differences in fitness the theory will remain intact.

Suppose that phenotype A is fitter than phenotype B and that each is equally common in the present population. Then, by the definition of fitness, the A descendants outnumber the B descendants. It then follows that, if the A and B phenotypes are sufficiently hereditary (i.e., if descendants are sufficiently similar to their ancestors), the A type itself outnumbers the B type, and there is a gradual increase in the proportion of the A type. The second empirical fact of natural selection is that *most phenotypes are sufficiently hereditary,* so that the numerous offspring of the fitter phenotype are themselves fitter. This should need no documentation, for we all have seen how like begets like. These two empirical facts are the foundation of natural selection. They show that fitter characteristics, when hereditary, are always replacing the less fit. Notice that the kind of heredity has not been specified. Characteristics passed from parent to child on chromosomes often are sufficiently hereditary, but so may be characteristics passed on through strong family tradition. It is clear from this that noninheritable characteristics play no part in natural selection for, although the bearer of one such trait may be fitter than that of another, its more numerous descendants will not have the trait. This is a perfectly adequate qualitative account of natural selection.

It used to be argued that natural selection was only a conjecture, because it had not been actually witnessed. By now we have become aware of many examples of natural selection in action. One of the best documented cases deals with the peppered moth (*Biston betularia*) which, within a century, changed from predominantly white to predominantly black, at least in certain areas. In Manchester, England, for example, the black form first became sufficiently frequent to be noticed in 1848. By about 1900 the black form had become dominant in Manchester and also in other areas, where industrial smoke had killed the lichens that had made the tree trunks whitish. Several other species of moths turned dark in other areas, where the tree trunks were darkened as a result of the industrial revolution. This phenomenon is called industrial melanism. Now, because both dark and light forms are still available, we can perform breeding experiments, and we find that the dark type is caused by a single dominant gene C. (This gene is not completely dominant, because the heterozygous Cc moths are less dark than the CC forms; actually, the Cc forms have become darker in the last 100 years, showing that C has become more completely dominant. We shall discuss it later as an example of the evolution of dominance.) This is a good example of natural selection: a change has unquestionably occurred, and it is a genetic change, as we have seen. It would be a perfect example of natural selection, if we knew why the black moths are fitter in industrial areas than elsewhere. The obvious guess is that the dark moths are protectively colored, hence are eaten less often, on the dark lichen-free bark in industrial areas, whereas the light moths are more difficult to see against the lichens. To test this hypothesis, Kettlewell (1956) released both dark and light moths, with inconspicuous paint marks under their wings, so that he could distinguish his released moths. He released them both in dark industrial woods and in the lichen-covered woods not polluted by industrial wastes. In successive days he recaptured as many as possible with light traps and by using captive virgin females to attract them. (In a control experiment both dark and light forms had been shown to be equally attracted by lights and by virgin females.) He recaptured significantly more light moths in the light wood and significantly more dark ones in the dark wood. The actual figures in one experiment are tabulated.

Moths	Dark Wood, Near Birmingham			Light Wood, in Dorset		
	Dark	Light	Total	Dark	Light	Total
Not recaptured	72	48	120	443	434	877
Recaptured	82	16	98	30	62	92
Total released	154	64	218	473	496	969

Kettlewell was able to show that the higher survival rate of the inconspicuous moths (i.e., the dark in the dark wood and the light in the light wood) was largely due to birds eating the more conspicuous ones.

As we shall see later, most species are well adapted, and it is the exception rather than the rule to find a situation like the industrial revolution with its consequent darkening of tree trunks, which causes spectacular changes in the species. More often the selection functions to maintain the status quo by eliminating phenotypically extreme gene combinations, which continually develop by gene recombination. Thus the mortality of babies in London hospitals caused a stabilizing selection. Figure 2-1 compares stabilizing and directed selection in a hypothetical situation.

Returning to our general account of natural selection, there are many additional questions we would like to ask. What is the source of new

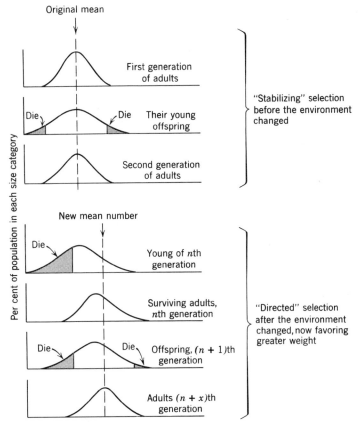

Fig. 2-1. A continuous phenotypic trait, such as weight, under the action of a hypothetical stabilizing selection (*top*) and directed selection (*bottom*).

traits, such as the dark coloration in the moths introduced into a population? Can we characterize the phenotypic effect of natural selection? (That is, can we name some trait which always changes in a prescribed direction under the action of natural selection?) Can we describe the rate of progress of natural selection? Is there any other means of evolution? These and other questions occupy the remainder of this chapter.

ON THE NATURE OF INHERITANCE

How important is the kind of inheritance geneticists study? Has it anything to do with intelligence, size, strength, or longevity, which affect fitness markedly, or does gene inheritance control only such minor characteristics as polydactyly and color blindness? It is difficult to answer these questions, because it can be maintained, for example, that the higher intelligence of children of intelligent parents is due to their intellectual environment and not to their chromosomes. The most valid way of distinguishing between the two is provided by twins. Identical twins come from the same fertilized egg and thus have identical chromosomes and genes. Fraternal twins are the result of separate fertilizations of separate eggs by separate sperm and are therefore as different genetically as two ordinary children in the family. Because the types of twins can be diagnosed in various ways—skin grafts from one person to another last only if the donor is an identical twin of the recipient, and fingerprints of identical twins are more similar than those of fraternal twins—we are in a position to say whether identical twins are more similar than fraternal twins. They are indeed in almost every respect. Why should this be? Often the parents do not know whether their twins are identical or fraternal, so that they do not provide different environments for different types of twins, at least not noticeably. Hence the degree of resemblance, which is so much greater in identical twins, must be due to the greater genetic similarity of identical twins. Such studies (Osborn and de George, 1959) show that probably more than one half the *difference* in these characteristics (intelligence, size, etc.) among people is hereditary. The fact that identical twins do differ shows, of course, that the environment, too, affects differences in these traits. Notice that it makes no sense to ask whether the trait itself is controlled by heredity or by environment, because in the absence of either of these conditions there could be no organism. We can ask the question only about differences among organisms.

Another kind of evidence is available about other animals and plants. When we breed corn, for example, we sort out the large ears from vigorous plants of one generation, from which to seed the next generation. Similarly, we can save for breeding only those cows that are themselves good milk producers. The success of these breeding schemes—the fact that the experimental corn plants are larger in successive generations, and that the milk-cow breeds produce more milk than their wild ancestors (even when the ancestral forms are grown in the modern fields)—shows that there is a hereditary component to these traits. The general conclusion from this kind of artificial selection is that there is a hereditary component to virtually every trait we can measure.

However, because it is difficult to distinguish individual gene effects in these traits of intelligence, fertility, and the like, we may still question whether the inheritance is really particulate (genes are the particles) or whether the offspring show a blending of parental traits, so that again the pure traits cannot be disentangled. Blending appears to be a reasonable assumption, because offspring are usually almost intermediate between their parents in these traits. However, blending cannot be the answer; instead, there is a large number of genes, each of which contributes a small amount to the quality of the trait. The simplest experimental demonstration is the following. When two pure-bred dogs of different breeds mate, the puppies are usually uniform. Blending could be accountable, but so could particulate inheritance, if we assume that the parents are homozygous because of long inbreeding. (Dog breeds breed true because they are highly homozygous.) Now, if we mate brother and sister puppies from this hybrid brood, we find a great variability among their offspring. According to blending inheritance, which is thus disproved, the puppies should differ no more than their parents did. Particulate inheritance, however, neatly accounts for this variability, as shown in the chart. For the F_1, uniform puppies would be heterozygous for the traits in which the parents differed. These heterozygous puppies, when mated, could produce, in their offspring, all combinations of the parental traits, as has been observed.

Breed A	Breed B	
Genotype: AA bb CC . . .	aa BB CC . . .	(each breeds true)
F_1 (puppies) genotype: all	Aa Bb CC . . .	(uniform)

F_2 (puppies of the F_1) genotypes:

AA BB CC	AA Bb CC	AA bb CC	
Aa BB CC	Aa Bb CC	Aa bb CC	(varied)
aa BB CC	aa Bb CC	aa bb CC	

Not only are there hereditary components to physical traits such as color, size, and intelligence, but there are also hereditary components to capacities to respond in various ways to the environment. For example, there is hereditary variability in the amount of heat required to cause a deformation during development in the fruit fly *Drosophila* (Waddington, 1957). Those that respond with little heat have offspring that generally respond with little heat. By selecting as parents for the next generation those flies that deformed with the least heat shock, Waddington even produced a strain of flies that deformed in the normal temperatures of their environment. Similarly, the degree of stunting of plants caused by wind and cold is hereditary, although the stunting itself is not necessarily hereditary. Clausen, Keck, and Hiesey (Clausen, 1951) showed this by transplanting forms of plants to new habitats. They found that different genotypes of the same kind of plant showed different degrees of stunting when transplanted to mountaintops, depending on how far removed they were from their original habitat; they all grew quite tall, however, when grown in their natural habitat. Other important characteristics, such as date of flowering, cold-hardiness, and luxuriance of vegetation also varied with genotype, and this variation was different in different transplant areas (Figs. 2-2 through 2-5).

It has even been suggested that this kind of heredity—the hereditary control of degree of response—can account for instincts. (Instincts, it

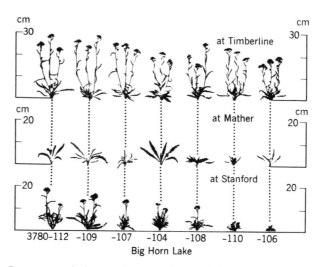

Fig. 2-2. Responses of clones of seven plants of the Big Horn Lake population from the crest of the Sierra Nevada (3350 m) at Stanford, Mather, and Timberline. The numbers at the bottom are those of the individual plants. (After Clausen, 1951). See Fig. 2-5 for location of stations.

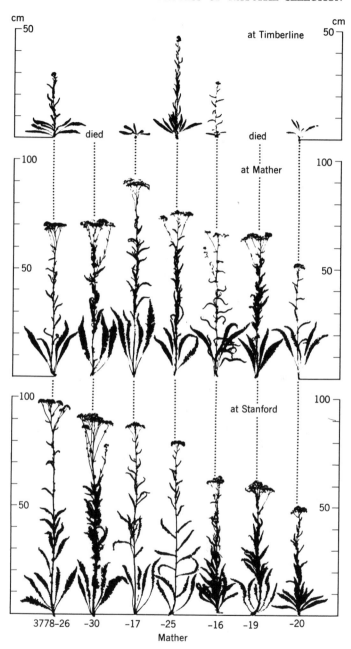

Fig. 2-3. Responses of clones of seven plants of the Mather population from the central Sierra Nevada (elevation 1400 m) at Stanford, Mather, and Timberline. The numbers at the bottom are those of the individual plants. (After Clausen, 1951.)

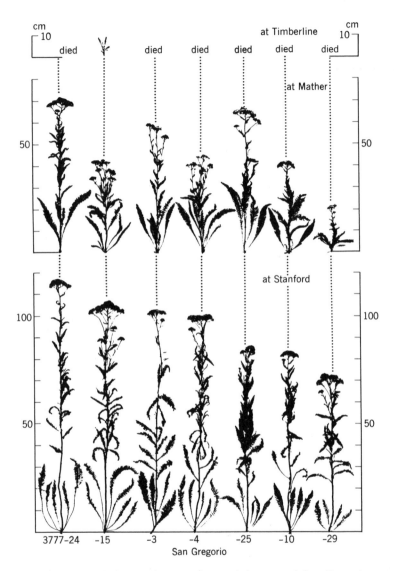

Fig. 2-4. Responses of clones of seven plants of the coastal San Gregorio population (elevation 50 m) at Stanford, Mather, and Timberline. The dotted lines connect herbarium specimens from members of the same clone at the three stations. The numbers at the bottom are those of the individual plants. (After Clausen, 1951.)

will be remembered, are behavior traits, which develop in an animal independently of normal variations in the environment and thus are, in a sense, hereditary.) In learned behavior the animal requires some conditioning before it responds properly. It must be punished for incorrect responses and rewarded for correct ones, until it has learned. Perhaps the amount of conditioning required before an animal responds is a hereditary trait. If so, as in Waddington's experiments already described, it may be possible to select animals that respond properly with less and less conditioning, until a strain is produced that responds without previous conditioning. Such a response would then be instinctive. (This "explanation" does not, however, tell us anything about how the instinctive information is stored or recalled, which are central questions in the study of animal behavior.)

SOURCES OF NEW VARIATION

Natural selection is often criticized on the grounds that selective destruction cannot be a creative process. Yet the trend of evolution from unicellular organisms to vertebrates and flowering plants has obviously been toward more and more complexity. Hence, so the argument runs, natural selection cannot be the mechanism of evolution.

An illustration is quite instructive at this point. Suppose that an army of monkeys were set to work randomly punching the keys of typewriters. As is well known, by chance alone, interspersed amid all the nonsense that they produced would be a few real words, which could be selected to construct any desired literary gem. To obtain the gem, strike out the nonsense! Of course, many thousands of pages of meaningless trash would be produced for each small piece of fine "creative" literature, but, given enough monkeys, we would be able to obtain copies of all the great creative literary masterpieces, *if* we only knew how to sort the literature out from the overwhelming quantity of nonsense. Thus we could be creative, with the aid of the randomly working monkeys—or machines—if only we had a method, such as natural selection, by which to remove selectively the trash and leave the masterpiece. Selective destruction, coupled with randomness, can thus be creative. This analogy shows how natural selection could indeed be creative, even if the source of new variability is random in direction, that is, if the type of new variation that appears is independent of the circumstance in which it appears. Of course, natural selection could operate if the new variation were appropriate to the situation, but the point is that it will also work if these variations are

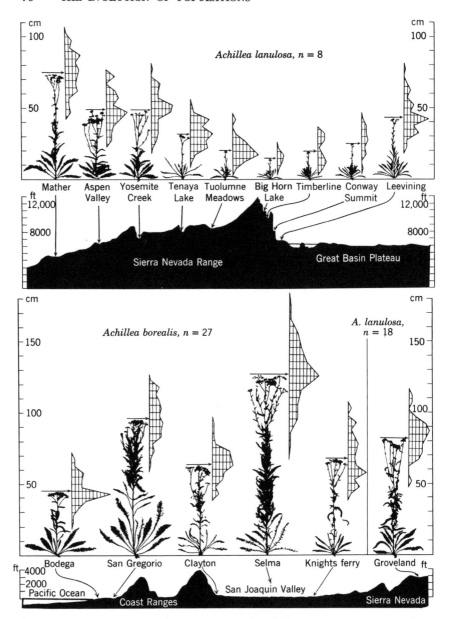

Fig. 2-5. Representatives of populations of *Achillea* as grown in a uniform garden at Stanford. They originated in the localities shown in the profile below of a transect across central California at approximately 38° north latitude.

Above: the western half of the transect, starting at the Pacific Coast; *below:* the eastern half, extending to the Great Basin plateau. The two parts together

random. (Random is used here in the sense of the typewriter—the key punched is not strongly dependent on the situation.) For brevity, we call the initiation of any hereditary novelty a *mutation*. Our result may then be restated: natural selection can be creative, even if mutations are not helpfully directed.

However, many biologists have claimed that mutations are helpfully directed, so that, for example, a mutation to heat resistance becomes likely in a hot climate, mutations to strong physique are more likely in a blacksmith, and mutation to pencillin resistance in bacteria results from exposure to penicillin. (Lamarck did not know about mutations, but this is a modern version of his well-known beliefs.) From the plant's or animal's point of view, such directed mutations would be a splendid achievement. This does not, however, indicate that they occur.

What evidence is there to show which view is correct? Are mutations helpfully directed or not? Most observed mutations are harmful, which suggests that mutations are not directed to be useful. There are two types of experiment which establish this fact with more certainty. The more recent and more spectacular one makes use of the "replica plating" technique due to J. and E. Lederberg (1952). The Lederbergs found that an exact replica of the pattern of colonies of bacteria in one petri dish could be transferred to another petri dish by pressing a taut velveteen disk into the petri dish containing the established culture and then pressing this disk onto a new plate. The stiff hairs pick up bacteria from the first plate and inoculate the second in the same pattern—hence the name replica plating. The Lederbergs then took a thriving culture of the bacterium *E. coli* and made a replica on a plate in which streptomycin had been added to the nutrient agar. As usually happens when bacteria are added to a medium which contains an antibiotic, a few scattered colonies appeared, which were composed of streptomycin-resistant bacteria. The question is, did the mutation to antibiotic resistance appear by random mutation at some earlier time in an ancestor of each of these resistant colonies or did the mutation to resistance occur as a result of the exposure? The Lederbergs

---

represent an airline distance of about 200 miles. Altitudes are to the scale shown in feet. Horizontal distances are not to scale.

The plants are herbarium specimens, each representing a population of approximately 60 individuals. The frequency diagrams show variation in height within each population: the horizontal lines separate class intervals of 5 cm according to the marginal scale, and distance between vertical lines represents two individuals. The numbers to the right of some frequency diagrams indicate the non-flowering plants. The specimens represent plants of average height, and the arrows point to mean heights. (After Clausen, 1951.)

found that by going to the site on the first culture (which was never exposed to the antibiotic)—the site corresponding in position to that occupied by a resistant colony in the second culture—they could find, after a suitable enrichment, streptomycin-resistant bacteria; that is, they could locate on the unexposed plate from which the replica was taken the brothers and sisters (so to speak) of the bacteria that, on exposure, had proved resistant. These "brothers and sisters" also proved resistant! Hence the mutation to resistance had occurred before the exposure in some ancestral bacterium. No adaptation to resistance occurred except in this way. Of course, it may be argued that the observed mutation was one that allowed the bacteria to become resistant on exposure and that the resistance itself is obtained only through exposure. This may be so, but it is quite irrelevant, because the experiment is set up to deal only with that mutation, which was the origin of the hereditary difference between the bacteria that were ancestors of resistant ones and those that were not. *That* mutation certainly arose independently of the exposure.

This result is very important. A second demonstration (historically earlier) will be described now. Luria and Delbruck (1943) conducted experiments which were further analyzed by Lea and Coulson (1949). Two separate results are to be compared here. In the first a broth of bacteria is stirred thoroughly and then plated onto agar in ten petri dishes. The agar has an antibiotic, although the broth has not. As usually happens, scattered colonies resistant to the antibiotic grew on many of the dishes. Nothing could be deduced from this fact alone. In the second part of the experiment broth was separated into ten parts and bacteria of common ancestry were grown for many generations in them. As before, each colony was then plated onto a petri dish with antibiotic in the agar. Again, scattered colonies of resistant bacteria appeared on the petri dishes, but there was more variability (fifty times as much) in the number of resistant colonies per petri dish. This experiment is hard to interpret, unless we consider what results we could expect of the two alternatives. If resistance appeared as a result of exposure to the antibiotic environment, the fact that in the two parts of the experiment there was a difference in the histories before exposure is not significant. The variability should be the same in the two parts, contrary to the observed result. On the other hand, if the mutation to resistance occurred randomly, independently of exposure, some of the ancestral bacteria in the broth cultures would have mutated to this form, and their descendants, too, would be resistant. In that part of the experiment in which the broth cultures were maintained separately sometimes an early mutation would result in a large number of descendants, all of the mutant form. In other broth cultures perhaps no mutations to resistance occurred. Thus there would be great variability in the number of mutants per culture. If, as in the first part of the experiment, the cultures were mixed before plating, this variability would be reduced. The experiment showed these results, thereby indicating that the mutation to resistance had occurred independently of the exposure. Of course, this does not show that a mutation is never directed by exposure, but only that these particular mutations are independent of exposure. As yet there is no contrary evidence.

The knowledge that the hereditary components of phenotypic traits are passed from generation to generation as developmental instructions coded onto DNA molecules makes this randomness of mutation direction plausible. Because all mechanisms make occasional errors, many mutations may be the errors in DNA duplication, these errors being reproduced in successive DNA replications. But "errors" normally are not helpfully directed.

This helps us to understand another important feature of mutations: they are repeated. Although a mutation may be a rare event, it is not unique, and others of the same type can be expected with a certain frequency (different rates for different mutations). Furthermore, mutations are reversible, so that if A mutates to B, B may also mutate back to A.

Nothing has been said about the rate of mutation being independent of the situation—only the direction has been shown to be independent or at least not appropriate. The rate indeed seems to vary strikingly with the circumstances. Thus there are mutator genes which increase the mutation rates of others. This makes possible an evolution of mutation rates, so that different mutation rates are found which are appropriate to their situations. H. J. Muller was given the Nobel Prize for demonstrating that X-irradiation of sex cells causes an increased number of mutations; many other artificial agents are now known to do the same. Although naturally occurring radiation doubtlessly causes some mutations, it cannot account for most of them. The usual cause of mutation is still not completely understood, but, whatever the cause, it does not seem to produce mutations that are particularly appropriate to the needs of the organism.

We shall now consider another common objection to natural selection. It is often pointed out that the mutations which geneticists have witnessed are almost always harmful. What makes us think, then, that enough beneficial mutations occur to account for all of the progress of evolution? We can immediately see one fallacy: because most mutations are reversible, if the mutation from $A \rightarrow B$ is harmful, that from

$B \rightarrow A$ must be beneficial. We can go even further. An ingenious argument due to R. A. Fisher (1958) shows that we should not expect visibly big mutations to be beneficial, although very small ones may often be. In essence, Fisher's argument is that if an animal is poorly adapted random mutations of visible size will sometimes improve the adaptation, but if an organism is nearly perfectly adapted, it will virtually never be improved by a big mutation. A microscope analogy is useful here: a random change in the focus of an out-of-focus microscope stands about half a chance of improving the focus. A random change in a well-focused microscope, however, cannot possibly improve the focus and will always be considered harmful. Thus we should expect only relatively small mutations to be beneficial; if an organism is already well-adapted, the mutation will have to be almost invisibly small to stand a chance of being beneficial. In view of this reasoning, the most logical answer to the question why visible mutations are nearly always harmful is that organisms are exceedingly well-adapted; only when there is a sharp change in the environment, which causes new phenotypes to be better adapted, do we expect a visible mutation to be beneficial occasionally. Such changes are, for example, the addition of an antibiotic to the medium of bacteria and the pollution of the environment by industrial wastes; as we have seen, it is only in such situations that we do indeed witness beneficial mutations.

The Quantitative Genetics of Populations. Because natural selection results in changes in the genetic composition of a population from generation to generation, it is important to have some numerical understanding of these changes. This will extend our qualitative picture of natural selection to a quantitative one. However, the results are not used again in this book.

For populations in which the selection takes place in the haploid stage, the quantitative description of change in genetic composition is simple. In fact, it is the description that we would use in describing the monetary investments on page 62. To set the stage for the diploid case, we shall present the mathematics in the form of a table, beginning with a population of 100 haploid individuals.

Table 2-1 is partly self-explanatory. The actual selective values (proportion of A, B genes surviving to reproduce) we write as SW_A, SW_B. We leave in it a multiplier, S, to show that it cancels out in the formula for the proportion of A genes in the next generation; that is, regardless of the values we place in this row of the table, as long as the *relative* selective values W_A, W_B are correct, we obtain the correct prediction for p' and p. Often we let $S = 1/W_A$ and obtain $W_A/W_A = 1$ and W_B/W_A for these two values. Thus suppose that the true values of the proportion surviving to reproduce are $W_A = 0.75$ and $W_B = 0.50$. Because $W_B/W_A = 0.50/0.75 = 0.667$, we could use 1 and 0.667 in the "proportion surviving" row just as well as the

Table 2-1

	Allele A	Allele B	Total
(a) Initial proportion	p	q	$p + q = 1$
(b) Initial number	$100\,p$	$100\,q$	100
(c) Selective value = actual proportion surviving to reproduce	SW_A	SW_B	(See comments in text)
(d) Number surviving to reproduce [(b) times (c)]	$100SpW_A$	$100SpW_B$	$100S(pW_A + qW_B) = 100S\overline{W}$ where $\overline{W} = pW_A + qW_B$

$p' = $ proportion A in next generation

$$p' = \frac{100SpW_A}{100S(pW_A + qW_B)} = \frac{pW_A}{pW_A + qW_B}$$

Change in p in one generation

$$\triangle p = \frac{pW_A - p(pW_A + qW_B)}{pW_A + qW_B} = \frac{pqW_A - pqW_B}{pW_A + qW_B}$$

$$\triangle p = (p' - p) = \frac{pq(W_A - W_B)}{pW_A + qW_B} = \frac{pq}{\overline{W}}\frac{\partial \overline{W}}{\partial p} \quad \text{(This line is for those who know calculus)}$$

actual values 0.75 and 0.50. In some of the later calculations this is not true.

In the diploid case mating preferences play a very influential role. If every individual insisted on mating with its own phenotype (an extreme case of "inbreeding"), there would soon be few heterozygotes left in the population. In fact, if heterozygotes were common, their matings would produce many homozygous recessives which, being unwilling to mate with the phenotypically dominant heterozygotes, would drain the recessive genes from the phenotypically dominant population. In similar manner, if each individual insisted on mating with a different phenotype (extreme "outbreeding"), progeny would be highly heterozygous. Random breeding is the dividing line separating inbreeding and outbreeding. It means that individuals show no mating preference for any phenotype and take what comes. It is easy to calculate the proportions of each diploid genotype in a randomly breeding population from the proportions of the genes themselves (at that locus) in the population. Before we do this, notice that the 3:1 or 1:2:1 and 1:1 ratios found in elementary genetics calculations of expected progeny of a particular cross do *not* refer to a whole population. In a whole population the different types of parental combinations occur with different, but predictable, frequencies, and each such parent combination has offspring whose proportions are governed by the familiar ratios. To obtain the proportions in the whole population, we must add the numbers of each genotype from each family. This is not so hard as it sounds if we realize that, in random mating, each gene in the gametes of one individual is equally likely to combine with any gene from the combined gametes of individuals of the opposite sex. In other words, *random mating of diploid individuals is equivalent in its effect on a pair of alleles to random union of gametes.* This simplifies matters enormously. If we call the fraction of A genes at one locus p and let $q = 1 - p$ be the remaining fraction of a alleles, each male A gene stands chance p of combining with a female gamete that has another A and chance q of combining with an a. Because fraction p of the male gametes are A, fraction $p \times p = p^2$ of the zygotes are AA. Similarly, fraction pq get a from the male and A from the female, so that a total of $2pq$ of the offspring are Aa. Finally, a fraction q^2 of the offspring are aa. Of course, $p^2 + 2pq + q^2 = (p + q)^2 = [p + (1 - p)]^2 = 1^2 = 1$, so that these fractions of the population which are of any given genotype add to unity, as they must. We can summarize this "Hardy-Weinberg formula" (named after its independent discoverers) as follows: if fraction p of alleles are A and fraction q are a in a large population and there is random breeding, the offspring in the whole population will consist of fraction p^2 of AA genotype, $2pq$ of Aa and fraction q^2 of aa genotype. Notice that p^2 and q^2 are the terms in the binomial expansion of $(p + q)^2$. (This is similar to saying that in 100 throws of a true coin 50 heads and 50 tails will appear; we know that for reasons of chance we should not expect precisely 50–50, but we do expect to observe a ratio near 50–50. Similarly, we expect to obtain only approximately the Hardy-Weinberg predictions.) Suppose, for example, that the recessive trait albinism aa appears in one person among 10,000. What is the fraction q of a alleles in the population? The formula would predict that a fraction q^2 of individuals should be aa; thus $q^2 = 1/10,000$, and therefore $q = 1/100$. In other words, the proportions of genes at this locus are about $p = 99/100$, $q = 1/100$. Applying the formula further, $2pq = 2 \times 99/10,000$, which is about $2/100$, of the individuals will be heterozygous. This means that 200 times as many individuals would have the a allele in heterozygous form as have it in homozygous form. It is for this reason, because heterozygotes show no phenotypic trace of albinism, that we feel justified in assuming that mating is random for such traits and that the formula is applicable. That is, although albinos may not be precisely equally desirable to normals as mates, the overwhelming proportion of a genes are in Aa form and, because they cannot be distinguished from AA, this proportion must unite randomly with A and other a genes. In what follows, we assume that mating is random.

Because each gene copies itself and because meiosis makes each allele equally likely to enter any given gamete, we can conclude that the proportions p and q will remain unchanged from generation to generation unless, of course, mutations, selection, or migration—or even random deaths—act to change these proportions. We now consider how recurring mutations and natural selection affect the frequencies p and q.

Let us begin by considering a special kind of selection. Suppose that recessive genotypes aa survive half as often as do Aa and AA, which have equal, perfect survival. For example, aa may have a coat color which is more conspicuous to predators. Suppose that there are 200 diploid individuals, with 400 genes at some locus, of which 200 are a and 200 A in a particular population at a given time. By natural selection the A genes will increase, of course. We proceed to calculate the change in p and q in one generation by means of Table 2-2.

Thus 350 genes survive, of which 200 are A and 150 a; the new frequency of A genes is 200/350. Because of the action of selection, A has then increased from $\frac{1}{2}$ to 200/350, that is, it has increased by 1/14, and a has been reduced from $\frac{1}{2}$ to 150/350. A second generation of selection will reduce this even further. Although the total number of genes has been reduced to 350 and the total number of individuals to 175 (notice that 25 aa died), we can assume that breeding will restore the population size to its original 200 and leave the new frequencies unaltered. (We are assuming that aa is not less fertile than AA or Aa.) Rather than pursue this special case further, we shall now derive a general formula for selection against a recessive in a population of n individuals, using the same steps and still assuming random mating. For concreteness, we assume that fraction s of the AA and Aa genotypes survive to reproduce, but that of the s surviving aa genotypes a proportion k fail to reproduce (i.e., $s(1 - k)$ succeed). Therefore the new proportion p' of A genes is (see Table 2-3)

$$p' = \frac{\text{number of surviving A genes}}{\text{number of surviving genes of either kind}}$$

$$= \frac{\text{number of surviving A genes}}{2\,(\text{number of surviving diploid individuals})}$$

$$= \frac{2snp}{2(sn - sknq^2)} = \frac{p}{1 - kq^2}$$

Notice that, as in the haploid case, the s cancels out.

Finally, the change $\Delta p = p' - p$ in proportion of A genes due to one generation of selection is

$$\Delta p = p' - p = \frac{p}{1 - kq^2} - p = \frac{p - p(1 - kq^2)}{1 - kq^2} = \frac{kpq^2}{1 - kq^2} \tag{1}$$

We obtain the total number of surviving individuals as follows: $snp^2 + 2snpq + s(1 - k)nq^2 = sn(p^2 + 2pq + q^2) - sknq^2 = sn(p + q)^2 - sknq^2 = sn - sknq^2$, because $p + q = 1$. We say that the $sknq^2$ individuals die "selective deaths," because they would have survived had their genotype been AA or Aa. They died because of the selection. Notice that, as in the haploid case, the s in numerator and denominator cancels in the Δp formula, so that we must only know the relative selective values to calculate Δp. When we deal with selective deaths, however, we must keep the s. We can easily test formula 1 by deriving the numerical example we have just studied. There $k = \frac{1}{2}$, $p = \frac{1}{2}$, $q = \frac{1}{2}$, so that, as we found before, by Equation 1

$$\Delta p = \frac{\frac{1}{2} \times \frac{1}{2} \times \frac{1}{4}}{1 - \frac{1}{2} \times \frac{1}{4}} = \frac{\frac{1}{16}}{\frac{7}{8}} = \frac{1}{14}$$

When k is small (i.e., when selection forces are weak) kpq^2 is small com-

Table 2-2

Initial frequencies: p(frequency A) = ½, q(frequency a) = ½

	Genotype AA	Genotype Aa	Genotype aa	Total
Proportion of zygotes	$p^2 = ¼$	$2pq = ½$	$q^2 = ¼$	1
Number of zygotes	¼(200) = 50	½(200) = 100	¼(200) = 50	200
Selective value; fraction which survive to reproduce	1	1	½	
Number of individuals which survive	50	100	25	175(2 × 175 = 350 genes survive) 25 died "selective deaths"
Number of surviving A genes	2 × 50 = 100	1 × 100 = 100	0	200
Number of surviving a genes	0	1 × 100 = 100	2 × 25 = 50	150 Total 350

Table 2-3

	Genotype AA	Genotype Aa	Genotype aa	Total
Initial frequency A: p; a: q				
Frequency of zygotes	p^2	$2pq$	q^2	1
Number of zygotes	np^2	$2npq$	nq^2	n
Selective value	s	s	$s(1 - k)$	
Number of surviving individuals	snp^2	$2snpq$	$s(1 - k)nq^2$	$sn - s\,knq^2$ survived $sknq^2$ die "selective deaths" (see text)
Number of surviving A genes	$2snp^2$	$2snpq$	0	$2snp(p + q) = 2snp$

pared with 1 so that the denominator of Equation 1 is virtually 1 and we can write the equation in the useful approximate form

$$\Delta p = kpq^2 \tag{2}$$

Applied to our numerical example, this approximation would predict $\Delta p = 1/16$ instead of the correct $1/14$, which is not too bad. Usually k is smaller than $\frac{1}{2}$ and formula 2 is correspondingly more accurate.

Selection does not always eliminate only recessives, of course, and we shall calculate the formula for Δp in the general case when all three genotypes have some failure to survive. However, before we do this, we shall consider some important theorems that are easier to prove for the special case of selection against a recessive gene. Because most harmful genes are recessive, this is the commonest situation, and restriction is only of slight concern to us.

First, suppose that, although selection is eliminating a, A is mutating to a in a fraction u of the A individuals in each generation. Then, while p (the proportion of A alleles) is small, selection is increasing p faster than mutation is reducing it, but there will be a level of p at which its increase due to selection is precisely balanced by its decrease, up, due to $A \rightarrow a$ mutation. At this value p_0, therefore, $\Delta p_{\text{selection}} = -\Delta p_{\text{mutation}}$ or, using approximate formula 2

$$kp_0q_0^2 = up_0 \quad \text{or} \quad u = kq_0^2 \tag{3}$$

We can solve this for $q_0 = \sqrt{u/k}$. Thus if we know either k or u, we can find the other in terms of q_0; if we know both, we can predict q_0. The hypothetical white robin phenotype, whose fitness was discussed on page 64, had a selective value only $\frac{1}{2}$ that of normals. Hence, assuming whiteness is recessive, $k = \frac{1}{2}$. If one robin in ten thousand is white, $q_0^2 = 1/10,000$. Therefore, mutation rate to whiteness u is $\frac{1}{2}(1/10,000)$ or $u = 1/20,000$, that is, one normal gene in 20,000 mutates to white per generation.

Of even more interest is the fact that the number of individuals killed in each generation because of having aa genotype, is $sknq^2$ and the proportion killed is $sknq^2/n = skq^2$. But our formula says that the mutation rate u also equals kq_0^2, so that, when mutation balances selection, $q = q_0$ and the number of individuals dying in each generation is equal to s times the mutation rate, *no matter what u is*. This is very remarkable. To reinforce our belief, we use the following heuristic reasoning: genes enter the population by mutation and leave it by selective death. Hence at equilibrium these rates must balance. If the mutant gene is relatively harmful, it is rare at equilibrium, and if it is only slightly harmful, it is commoner at equilibrium. A large fraction of the rare, harmful gene dying eliminates the same fraction of the whole population as the small fraction of the common, less harmful gene. As an example, suppose we know that a certain amount of hydrogen-bomb testing will increase radioactivity and that the mutation rate will double. Then we can predict that the number of deaths due to homogyzous *recessive* mutants will also double, regardless of whether the mutations are

lethal in homozygous form or are only slightly harmful. Because most mutations are recessive to wild type, this is probably a true generalization.

Let us now turn to the question how long it takes selection to change
gene frequency by a given amount. Suppose that our population is almost
all *aa*, when a dominant *A* gene, which entered by immigration or mutation,
becomes favorable because of a change in the environment. How long will
it take *A* virtually to replace *a*? Notice that, as *a* gets rare, it appears more
and more often in *Aa* form, in which it is concealed. Hence it becomes more
and more difficult to reduce *a*, as *a* becomes scarce. In fact, when there is
but one *a* gene left, it must combine with an *A* in heterozygous form so that
selection will be powerless to remove it. (Chance may, however). Let us
ask how many generations it will take to reduce the frequency *q* of *a* to
5 per cent, given that initially *q* was 95 per cent (the other 5 per cent being
the mutant or immigrant *A* genes). We have formulas 2 and 3 for the change
in gene frequency per unit time; to find the time required per unit change
in gene frequency, we simply invert this formula. More precisely, we must
integrate as follows:

$$\text{time} = \int_{q=0.95}^{q=0.05} dt = \int_{p=0.05}^{p=0.95} \left(\frac{1}{dp/dt}\right) dp$$

because *p* goes from 0.05 to 0.95 as *q* goes from 0.95 to 0.05.
As Δp is the change in *p* per generation, Δp is approximately equal to dp/dt,
where *t* is measured in generations.

We can now substitute this in our formula for time:

$$\text{time} = \int_{0.05}^{0.95} \frac{dp}{kpq^2} = \int_{0.05}^{0.95} \frac{dp}{kp(1-p)^2} = \frac{1}{k}\left[\frac{1}{1-p} + \log_e \frac{p}{1-p}\right]_{0.05}^{0.95}$$

$$= \frac{1}{k}\left[\frac{1}{1-0.95} + \log_e \frac{0.95}{1-0.95} - \frac{1}{1-0.05} - \log_e \frac{0.05}{1-0.05}\right]$$

$$= \frac{1}{k}\left[20 + 2.94 - 1.05 + 2.94\right] = \frac{26.93}{k} \tag{4}$$

Therefore the change requires about $26.93/k$ generations. For example, if *aa*
is only slightly less fit than *Aa* and *AA*, so that, say, 99/100 as many *aa*
survive as do the others, $k = 0.01$ and the change will take 2693 generations. If, on the other hand, only 90 per cent as many *aa* as *Aa* or *AA*
survive, $k = 0.1$ and only 269 generations are required. We could, of course,
put in any initial and final frequencies in place of the 0.05 and 0.95 in
formula 4. For example, to go from $p = 0.01$ to $p = 0.99$ would take
$110.2/k$ generations. Eliminating a dominant *B*, for example, is, of course,
much quicker once the *bb* genotypes are common enough to be frequent.
But this process is slow-starting because the fitter *b* genes are hidden by
B's in heterozygous form. Genes of intermediate dominance, in which
heterozygotes can be differentiated from either homozygote, are the fastest

ones to substitute. We can take an example of the use of formula 4 from Kettlewell's experiments cited on page 68. In the dark woods $82/154 = 53.2$ per cent of the dark moths were recaptured and $16/64 = 25$ per cent of the light moths were recaptured. Hence each light moth had $25/53.2 = 0.47$ as much chance of surviving. Thus the *relative* fitnesses of CC, Cc, and cc were 1, 1, 0.47, although we do not know the actual fitnesses. Because $0.47 = 1 - 0.53$, the k of formula 4 is 0.53, and by (4) we expect the change from 5 per cent C gene to 95 per cent C to have taken $26.93/0.53 = 49.7$ generations, which is of the right order of magnitude, suggesting that selection is responsible for the observed changes.

Finally, let us derive a more general formula for the change in gene frequency under the action of natural selection. Because we want to allow genes to be intermediate in dominance, we call the alleles A_1 and A_2, instead of A_1a. If the initial frequency of A_1 is p and A_2 is q

Table 2-4

	Genotype A_1A_1	Genotype A_1A_2	Genotype A_2A_2	Total
Frequency	p^2	$2pq$	q^2	1
Number	np^2	$2npq$	nq^2	n
Relative selective value	1	$1 - \delta k$	$1 - k$	
Relative number surviving	np^2	$2n(1 - \delta k)pq$	$n(1 - k)q^2$	$n[1 - kq^2 - 2\delta kpq]$ individuals, two times as many genes

δ measures degree of dominance. If $\delta = 1$, A_2 is dominant; if $\delta = 0$, A_1 is dominant.

$$p' = \text{new frequency of } A_1 \text{ genes} = \frac{2n[p^2 + (1 - \delta k)pq]}{2n[1 - kq^2 - 2\delta kpq]} = \frac{p[1 - \delta kq]}{1 - kq^2 - 2\delta kpq}$$

$$\Delta p = p' - p = \frac{p[1 - \delta kq]}{1 - kq^2 - 2\delta kpq} - \frac{p[1 - kq^2 - 2\delta kpq]}{[1 - kq^2 - 2\delta kpq]}$$

$$= \frac{pqk[q + 2\delta p - \delta]}{1 - kq^2 - 2\delta kpq} \tag{5}$$

Now notice that the mean relative survival rate, which we call $\overline{W}$, is

$$\overline{W} = \frac{\text{total number surviving}}{\text{total number}} = \frac{n[1 - kq^2 - 2\delta kpq]}{n} = 1 - kq^2 - 2\delta kpq$$

Then, because $q = 1 - p$,

$$\frac{\partial \overline{W}}{\partial p} = 2k(q + 2\delta p - \delta)$$

Substituting this in formula 5, we obtain

$$\frac{dp}{dt} = \Delta p = \frac{pq}{2\overline{W}} \frac{\partial \overline{W}}{\partial p} \tag{6}$$

which, except for the 2 (due to diploidy) is identical to the equation for Δp on page 81 for haploid selection. This is a very general formula for gene frequency changes under the action of selection. Its principle virtue is that $pq/2\overline{W}$ is always positive. Hence Δp is positive when, and only when, $\partial\overline{W}/\partial p$ is positive. In other words, if an increase in p increased $\overline{W}$ (i.e., $\partial\overline{W}/\partial p > 0$), p would increase; if a decrease in p increased $\overline{W}$ (i.e., $\partial\overline{W}/\partial p < 0$), p would decrease—p does whatever will increase $\overline{W}$. If we are given $\overline{W}$ as a function of p, as in Fig. 2-6, we can interpret this more easily.

If p is now X on the figure, we can predict that it will decrease to point Y, at which time no further small changes in p will increase $\overline{W}$. If there happens to be a point Z with higher value of $\overline{W}$, natural selection will not reach it without some different kind of force acting. Perhaps chance processes will achieve the transition to the main "hill" in the figure. Conceivably, even a change in the degree of inbreeding may, because all our formulas refer to random breeding. $\overline{W}$ is a different function of p when there is inbreeding or outbreeding, and perhaps this function has its highest point atop the hill on which p already is. This is a picturesque way of viewing the effects of natural selection, but it must be used with extreme caution, always remembering that the function $\overline{W}$ will vary greatly from season to season and place to place and will also vary, as the degree of inbreeding changes or as gene frequencies at other loci change.

A second virtue of Equation 6 is that $\Delta p = 0$ if and only if either $\partial\overline{W}/\partial p = 0$, or $p = 0$ or $q = 0$, that is, if and only if $p = 0$ or $q = 0$, or

$$(1 - 2p)\delta = q = 1 - p$$

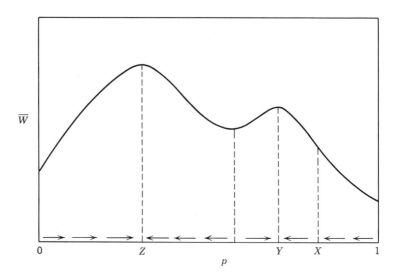

Fig. 2-6.   Mean selective value $\overline{W}$ of whole population plotted against gene frequency, p. Arrows show how natural selection would change gene frequency.

that is

$$p = \frac{1 - \delta}{1 - 2\delta}$$

If δ is positive, this requires $p > 1$, which is impossible. However, if δ is negative, meaning that the heterozygote survives better than either homozygote, there is a value of p between zero and one for which there is equilibrium. This situation is discussed further in the next section.

At this point it is appropriate to discuss the principal shortcomings of this quantitative theory of natural selection. Basically, in order to predict the course of evolution, we must know how phenotype F changes with time. In other words, we must know dF/dt. If phenotype is a function of any variables X_i, where $i = 1, 2, 3, \ldots ,$

$$\frac{dF}{dt} = \sum_i \frac{\partial F}{\partial X_i} \frac{dX_i}{dt}$$

Wright (1931) gave the relation between gene frequency X at locus i and time in terms of $d\overline{W} / dX_i$, as we have already seen. To obtain dF / dt from this, we write

$$\frac{dF}{dt} = \sum_i \frac{\partial F}{\partial Xi} \frac{dX_i}{dt} = \sum_i \frac{\partial F}{\partial X_i} \frac{X_i(1 - X_i)}{2\overline{W}} \frac{d\overline{W}}{dX_i}$$

Here we have two unknowns: we must learn the relation between F and X_i and the relation between $\overline{W}$ and X_i. In other words, the easy, deductive relations are worked out, but the hard, empirical ones remain for future investigators. Only in a few cases do we know the precise relation between phenotype and fitness, between mutation rate and $\overline{W}$, or even between phenotype and genotype. It is in investigations of these relations that we expect future progress.

MAINTENANCE OF VARIABILITY

We have studied the sources of hereditary variability, which we call mutations, and we have seen how fast the genetic composition of the population changes toward the final state. How then do we account for the fact that most populations are always variable? From everyday experience we know that the population of people alive now has about the same variability as it had ten years ago and as, we infer, it had one thousand or ten thousand years ago. We still see tall and short, fair and dark, intelligent and stupid people. Nevertheless, we would expect that natural selection would quickly sift out the less fit forms and leave a uniform population.

Actually, there are several methods of storing genetic variability and preventing natural selection from eliminating all but one allele at each locus, and geneticists are by no means agreed on which of these methods is used under which circumstances.

One of the most important features of recurring mutations is their ability to balance the action of natural selection. We have already seen this for recessive mutations on page 79. More generally, if one gene A in one thousand mutates to allele B, in a population of one hundred thousand individuals the number of new B mutants per generation will vary from zero to 100, as the proportion of individuals which are A varies from zero to one. In fact, if the proportion A is p, the number of new B mutants will be $100,000p/1000$ (see Fig. 2-7). However, if natural selection and back mutation from B to A are eliminating B genes, the number of B genes eliminated in each generation will decrease to zero as the proportion of A increases to one, although the exact shape of the curve will vary according to the situation (i.e., according to whether A or B is dominant, etc.) Hence the two curves of Fig. 2-7 intersect above some point p', representing the proportion of allele A in which the rate of elimination of A genes precisely balances that of B genes. This is a stable balance, too, for, if the proportion falls below p', selection and back mutation will increase the A genes at a faster rate than that at which mutation of A to B is occurring, thus

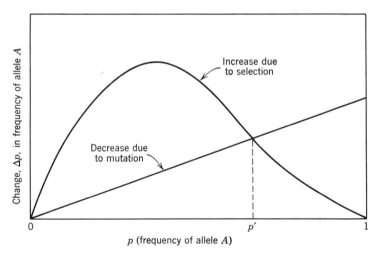

Fig. 2-7.   The decrease, due to mutation away from allele A, and the increase in A due to selection are plotted in the same figure for a sample situation. Where the curves intersect (at p'), the decrease equals the increase and there is no further change. When $p > p'$, it will decrease, and when $p < p'$ it will increase; hence p' is stable.

restoring the balance; if the proportion increases above p', it will fall again for similar reasons. For example, the gene that causes hemophilia (a sex-linked gene causing bearers to bleed freely when cut) seems to be maintained in human populations by a low-mutation rate, thus opposing the elimination of the gene by natural selection. The fitness of "bleeders" is very small, but the frequency of the gene remains about the same.

There are other ways of maintaining genetic diversity. When a heterozygous genotype AB is fitter than either of the homozygotes AA or BB, neither A nor B can be eliminated (we proved this in the last section). Heuristically, the reasoning is as follows: if either A or B should become rare, it will then most likely combine as a gamete with a gamete containing the other allele and appear as a zygote in heterozygous form, which is the fittest. Therefore, whichever gene becomes rare becomes also heterozygous, hence fitter, and is thus restored to its proper abundance. The most picturesque example of this situation, usually called balanced polymorphism, is the sickle-cell anemia gene, which in homozygous form usually causes death, by anemia, in childhood. However, there is evidence (Allison, 1956) that in the heterozygous form the gene confers some protection against a kind of malaria. When malaria is prevalent, as in East Africa, both homozygotes are likely to die: one from malaria and the other from anemia. Hence the heterozygotes are the fittest, and therefore both genes persist. This accounts for the high incidence of the sickle-cell gene in African populations. If the gene were maintained by recurrent mutations, we would expect Americans of African descent to have the same high frequency, but they do not. This is readily explained by the fact that malaria has been virtually eliminated in America, so that the heterozygote has little, if any, advantage.

Somewhat similarly to the heterozygote superiority, if the environment contains patches of different types, such that AA is fittest in one patch and BB is fittest in another, selection may act independently in the two patches and may not completely eliminate either. No one knows how important this means of preserving variability is, because no one has found a good way of distinguishing it experimentally from other possibilities.

THE COST OF NATURAL SELECTION

If we turn our attention from the survivors of natural selection to those who died because of the action of natural selection, we discover a very neat way of picturing genetic changes. The distinguished British biologist J. B. S. Haldane was the first person to do this. Haldane (1957) defined a selective death as one that would not have occurred if the dying organism had had the best genotype. Thus the $snkq^2$ individuals, which were removed from each generation in Table 2-3 on page 84, were the selective deaths. Extra births that would have occurred if the organism had had the best genotype are properly included. That is, selective "deaths" need not be real deaths—they can simply be a lower reproductive rate (as might be induced by taking monastic orders).

Haldane asked himself how many selective deaths occur during the whole course of the replacement of one gene by a superior allele. He obtained the striking answer that almost thirty selective deaths per individual are required, no matter how superior the new allele is. We shall give proof of it, but first let us consider some of its consequences. Basically, it permits us to estimate the numbers of genes substituted in the course of natural selection, even though we know nothing about the comparison of the alleles involved in the substitution. Thus, although we seldom know whether the new gene is fitter by one-tenth or by one one-thousandth, we can still say that about thirty selective deaths per individual will accomplish the gene substitution. If we are willing to take the risk and to estimate that, for example, an average of one-tenth of a selective death per individual per generation has occurred during the normal course of evolution, *we can conclude that about one gene will be substituted in every three hundred generations.* (Three hundred because 300 times $\frac{1}{10}$ equals 30, the required number). Notice that the selective deaths need not be used up on one gene substitution after another. Many genes can be substituted simultaneously; that is, one quite superior gene, using a large number of selective deaths per generation, may be substituted quickly, or several genes that are only slightly superior, each using only a fraction of the selective deaths, may be substituted slowly but simultaneously. In either case an average of about one substitution per three hundred generations will be accomplished. The vagueness of this number 30, which we are using as a rough rule of thumb, stems from the fact that, although the number is virtually independent of the superiority of the new gene, it does depend on how dominant it is (i.e., on how intermediate in fitness the heterozygote is) and, of course, on the frequencies of the new and old genes before and after substitution as well as on other factors, such as gene interaction. Also, it assumes that the population size remains approximately constant, which is a serious drawback. Most evolution probably occurs around episodes of colonization and population fluctuation. Thus 30 is only an educated guess of what the average value might be.

A more serious drawback of this kind of calculation is its assumption that the population has a fixed capacity for selective deaths. It is true that a population has a certain capacity to tolerate a proportion of deaths in each generation, but it is by no means clear that these deaths are selective. This only serves to set an upper limit to the number of selective deaths. Furthermore, selective deaths need not be a burden to the population, for, when a distinctly beneficial mutation becomes established, its bearers will survive no matter how many of the bearers of the allele perish. Bacteria in an antibiotic medium must suffer nearly 100 per cent selective mortality, with only the very few resistant mutants surviving. Furthermore, when selection is so strong that most organisms die selective deaths, the estimate of 30 is no longer valid. For example, if the environment changes so that the dominant gene is lethal, one selective death per dominant phenotype will accomplish the elimination. However, these violent changes are probably very rare compared with slow, unspectacular ones during the course of evolution, so that we shall continue to use Haldane's useful estimate, but we shall use it with caution.

There is one important situation that leads to more rapid gene substitution than the 300 generations we are using as a rough guess. When a single pair of individuals from a mainland population starts a new colony on an island, the whole colony is descended from that pair. If the pair happens, for example, to be homozygous for some rare gene at some locus on some chromosome, the whole colony of their descendants will also have this gene. Here a gene substitution has been accomplished in one generation, which in the mainland population would have taken about 300 generations.

For the reader familiar with calculus we shall now give proof of a version of Haldane's result, assuming selection against a recessive gene. An understanding of this proof is not necessary for the rest of this book. We shall consider randomly breeding diploids. As already noted, the number of selective deaths per generation, when selecting against a recessive whose homozygotes have selective value $s(1 - k)$ compared with s for the dominant phenotypes, is $snkq^2$. We want to add these deaths over all the generations required to accomplish the substitution of one gene for another. That is, the total number of selective deaths, D, which take place while the dominant gene rises from proportion p_0 to proportion p_1, is given by

$$D = \int_{p=p_0}^{p=p_1} snkq^2 \, dt = sn \int_{p_0}^{p_1} \frac{kq^2 \, dp}{dp/dt}$$

Now, by Equation 2, page 85, $dp/dt = kpq^2$, very nearly, and is independent of s. Thus

$$D = sn \int_{p_0}^{p_1} \frac{kq^2}{kpq^2} \bigg| dp = sn \int_{p_0}^{p_1} \frac{dp}{p} = sn \, [\log_e p_1 - \log_e p_0]$$

If p_1 is near 1, $\log_e p_1$ is near zero and can be neglected. Hence approximately

$$D = -sn \log_e p_0$$

Thus, if p_0 is $1/1000$, 6.9 sn deaths will accomplish the change. If $p_0 = 1/10,000$, 9.2 sn deaths are required, etc. More generally, the heterozygotes are intermediate, hence also suffer some selective deaths. With the typical degree of dominance, Haldane figured that about three times as many deaths would be required as in selection against recessives, thus raising the figure to about 30 sn. When s is very different from 1, the population size will change radically, because all genotypes are being altered in this proportion. Hence in a relatively stable-sized population s will be near 1, and the formula reduces to about 30 n.

GROUP SELECTION

Perhaps the biggest unsolved problem of natural selection—the problem that more than any other makes evolutionists get angry and say something irrational—is concerned with whether the kind of natural selection we have considered is the only one possible. We have measured the fitness of a character in terms of whether its bearer leaves descendants. According to these terms, for example, an altruistic individual who endangers his own life to save others not related to him must be less fit, and his type will be eliminated by natural selection.

In his beautifully written book *Evolutionary Theory and Christian Belief,* Lack even cites this generosity as evidence of the existence of God. For, he says, man is altruistic. And, he goes on, evolution, the only known cause of biological properties, cannot account for altruism. Hence some supernatural Being must have given it to man.

Notice that altruism practiced on one's own young can be easily accounted for, because it is fitter to risk one's life a little to save one's own offspring. It is practicing altriusm on strangers that must still be accounted for. No evolutionist doubts that ordinary natural selection eliminates this kind of altruism, but there are many who feel that a kind of "group selection" also occurs in which more harmonious groups of organisms (whether they be societies, species, or whole communities) replace less harmonious ones. These people argue that, although the altruist is endangering his life and that of his descendants, he is making his society safer so that it will survive while a society without such altruists would perish. This is theoretically possible if nonaltruistic societies are replaced by altruistic ones faster than the altruistic genes are eliminated by ordinary natural selection within the populations. We must also require that newly founded colonies be purely altruistic. However, the extent to which this "group selection" has molded present-day life is completely unknown, and surprisingly few people are investigating it. We shall discuss it further in Chapter 4, which discusses social behavior. Perhaps a plausible case of group adaptation is the adaptation of parasites not to kill their hosts. This case at least serves to point out the difficulty of classifying selection into "ordinary" and "group" selection.

3

The Results of Natural Selection: Species and Other Adaptations

SPECIES

Names for objects are an essential part of our language. When we speak of an animal often, we attach a name to that particular individual (a pet, for example). However, it would burden our vocabulary too much to have a separate name for all of the billions of individual animals and plants. Instead, we try to distinguish sets containing organisms which are suitably similar and to assign the same name to all of them (e.g., measles virus, maple tree, robin). The usefulness of such a name is greatly enhanced if the set of organisms to which it refers is cohesive and discrete, that is, if all are similar to others in the set but different from all other organisms. For example, the name *robin* is particularly useful, because all robins resemble one another more than any one of them resembles any other kind of bird. We do not normally bother to attach distinct names to sets which are not discrete (colors are an exception). Although we have names for several discrete kinds of maple tree, we have no separate names for large and small maple trees, because they come in all sizes. This motivates a consideration of cohesive, discrete sets of organisms, for which it is especially appropriate to have names.

First, we shall use our knowledge of natural selection to predict when discrete, cohesive units should occur. We know that selection favors phenotypes that leave more descendants, and, in most environments, they are the phenotypes that make the most effective use of their resources, other things being equal. Hence we begin by

96

asking when two discrete populations are, in combination, more efficient than a single uniform one at utilizing the available resources. The answer is relatively easy, if we first classify all environments into two categories, which we call "fine-grained" and "coarse-grained" relative to the organism under discussion. For simplicity we picture the environment as a patchwork of two kinds of resources. We call this "fine-grained" if, in its daily activities, our organism consumes these resources in the proportion in which they occur. If the patches are sufficiently large, so that the organism can, and does, select and use one in preference to the other, it is coarse-grained for that organism. For example, the canopy of an oak-hickory forest may be fine-grained for a scarlet tanager, which forages indiscriminately in both oaks and hickories, but coarse-grained for the defoliating insects, which attack oaks or hickories preferentially. For individuals in a coarse-grained environment, the answer to our question is immediate. Because the individuals can spend their time in patches of their choice, discrete populations are more effective whenever (and this is virtually always) the different resources can be better utilized by different adaptations—a jack-of-all-trades is a master of none. For fine-grained resources the answer is still elementary, but it is somewhat more complicated, because several possibilities, which were irrelevant in the coarse-grained case, are now important.

We begin the fine-grained case by an analysis of the one situation in which a single jack-of-all-trades is actually more effective than a combination of two specialists. In all other cases two discrete populations will be superior, and we shall discuss them later. Suppose, first, that the two resources are very similar, so that a phenotype which is good at utilizing one is at least moderately good at the other. An example is graphed in Fig. 3-1, in which the abscissa is any continuous phenotype coordinate—for example, bill length in a bird, for concreteness. The two solid curves are the effectiveness in utilizing the two resources separately. Thus the bill length S_1 (S for specialist) is optimal on resource 1 alone and bill length S_2 is optimal for resource 2 alone. If, however, our environment is a fine-grained mixture of equal parts of resource 1 and resource 2, we plot a dotted line at a height just halfway between the heights of the two solid lines, indicating the effectiveness on the mixture. Notice that the jack-of-both-trades, J, proves superior to either specialist and, it can be proved, to both specialists in combination (MacArthur and Levins 1964). Under this condition it would be desirable to have one population, not two, on these resources. Suppose, now, that the resources were less similar, as indicated by the solid curves in Fig. 3-2. These curves are

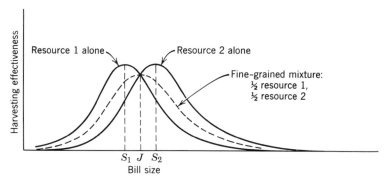

Fig. 3-1. Harvesting efficiency is plotted against a sample phenotypic trait for two similar pure resources (solid curves) when occurring separately, and for a fine-grained mixture of the two (dotted line). The S_1 and S_2 refer to specialist phenotypes, and J to a jack-of-both-trades.

sufficiently separate, so that the combined dotted curve for the mixed resources has two peaks separated by a trough in which the jack-of-both-trades lies. Here the two specialists are clearly superior to the intermediate phenotype, J. Furthermore, the two figures show that there is likely to be a limiting similarity to two coexisting species: when the resources and optimal phenotypes reach a threshold similarity, the two are replaced by one intermediate population.

In summary, we have shown that two discrete populations are superior to a single *uniform* one, if they depend either on coarse-grained resources or on the rather different fine-grained resources. Only if the resources are fine-grained and quite similar is a single uniform population theoretically superior.

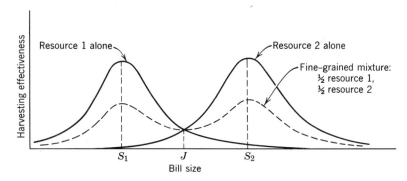

Fig. 3-2. Same as Fig. 3-1, except that the two resources are now quite different, so that the dotted line (the average of the two solid curves) now has two humps. In the mixed environment, J is now inferior.

Before turning to the question whether these theoretical optima are actually achievable, we complete the answer to the first question. We still have not shown when two discrete populations are superior to one extremely *varied* population. There are three principal reasons why a varied population is not the appropriate answer for adaptation to patches of different, perennially available resources. (For a more sophisticated analysis, which includes temporally varying resources, see Levins, 1962). The first is that parents often need to train their young, and in an extremely variable population the young might be sufficiently different so that the parents' training would be inappropriate. Second, the laws of population genetics rather than the availability of resources would control the relative abundance of the variants, which is less efficient. Finally, the genes of discrete, rather uniform populations seem to be quite well attuned to one another in cooperative "coadapted" complexes. A population with great variability would be perpetually breaking up these complexes.

Hence we have seen that under certain circumstances—different, or coarse-grained resources—discrete populations would be an optimal strategy. But is this strategy attainable? Empirically, the answer is often, but not always, yes.

If we collected all birds on a small island and examined them, we would find that we could assemble them in little piles in such a way that all of the individuals in each pile would be strikingly similar to one another (although not identical) and that no individual in any pile would be very similar to any individual in any other pile, each pile being separated from the others by a distinct gap. For convenience in referring to the different types of birds, we would attach a name, such as "robin," to the set of individuals of each of these discrete piles.

We might discover, on dissecting some individuals, that some piles consisted wholly of females and others of males. Observations on their life histories would enable us to associate the females with the appropriate males and to give them both the same name. These observations would do even more: they would show us that, almost without exception, the males given one name (e.g., robin) breed only with the females of the same name and leave offspring of the same name. Thus the names would refer not only to individuals that look alike but also to individuals, all of whom, at least theoretically, could share a common recent ancestry. If someone else were to try the same experiment, he would almost certainly end up with the same individuals associated into piles (although, of course, he might give them different names). In fact, in primitive tribes that depend on hunting and gathering for a living, hence where classification is a vital occupation, it is gratifying to find that the natives have attached names to the same sets of individuals as have visiting scientists.

Individuals that share a name, and probably an ancestry, have been called "species," and there is no reason to doubt that they act as natural, discrete units—provided that we keep our island small and limit our sample to sexually reproducing organisms like birds. Moreover, whatever modern definition we give to the word "species," it must be such that these natural units correspond to the defined ones.

Before we provide a definition, however, it is worth considering a few other examples which do not function so well. Suppose, instead, that we collected all of the irises (or some other plant usually reproducing asexually) on the island and attempted to arrange them in piles as we did the birds. It would seem easy at first, and we would begin arranging them with confidence. Soon, however, the amount of variability among extreme plants within a pile would greatly exceed the difference among the more similar plants in different piles, and the piles would become arbitrary. If we tried to prevent this, and arranged each clone of plants in a separate pile, we would have far too many piles to name conveniently. It is to be remembered that the names were issued to make future reference more convenient. We did not find many birds deviating from the rest of their pile, for, if one deviant should arise, it would either mate with more normal individuals and have, by the rules of genetics, offspring that were less extreme, or more probably it would not mate at all, being too different to be recognizable as a potential mate by the other members. An extreme iris, not being involved in sexual reproduction, could leave extreme descendants without difficulty. Now, although it is still necessary for our convenience to name irises, and although the names are often called species names, the groups so named do not behave as a natural unit, and for the purposes of this book we shall use the word "species" for the sexually reproducing, natural units which were described for the birds. We will also exclude self-fertilizing sexual organisms, such as many plants, which behave as asexuals in not having any cohesion because of interchange of genetic material.

Suppose that we were to collect all the birds from the whole world. Aside from the uncertainty of catching them all, we would also find it difficult again to assign them to piles. Often a species, which seems perfectly good in one locality, changes gradually as we proceed across a continent, until eventually it is too different to be placed in the same pile. As in the case of the iris plants, the extreme individuals in a pile are not eliminated, but for a different reason: although sexual reproduction exists, it does not take place between birds which return each year to opposite sides of a continent. Thus, just as we had to eliminate asexual organisms from the collection of individuals that could be divided into natural species, also must we eliminate collections combined from very different regions. We cannot inquire whether "robins" now in California are the same species as or different from the "robins" now in New York. The name "robin," attached for our convenience, is used for both, because they change gradually and are really quite similar.

Finally, if we examined a sequence of fossils (of complete specimens, if possible) taken from one place but extending over millions of years, we would have difficulties analogous to those that we had with the collection taken at one time but extended widely over space. At any one time, the birds can be neatly assigned to piles, but, as they change gradually with time, they eventually become so different that they cannot be assigned to the same pile as their original ancestors. Thus we cannot divide into species a collection of specimens taken at very different times.

For the remaining individuals—those that are sexually reproducing, are not self-fertilizing, live in one place, and are contemporary—we can make a satisfactory species definition: *species are the discrete sets of individuals whose conformity is assured by regular exchange of genetic material in sexual reproduction.* It is difficult to say how frequent the exchange need be, in practice.

(It is frequently possible to determine whether two individuals belong to different species by observing whether they breed and have fertile offspring. However, we must be cautious in interpreting such observations; mallard and pintail ducks, which are perfectly distinct species, readily breed in captivity and form varied and generally intermediate populations. Apparently, some of the barriers that prevent frequent interbreeding in the wild are not present in the artificial circumstances of zoos and parks.)

Many organisms are usually asexual but also require occasional sexual reproduction. Whether "species" can be defined in such organisms depends on whether the sexual reproduction is sufficiently frequent to ensure the cohesion of interbreeding individuals into a unit.

Mathematicians call a relationship like "A is the same species as B" between individuals an "equivalence relation," if for all individuals A, B, C, etc., (1) "A is the same species as A" and (2) "A is the same species as B" implies that "B is the same species as A." (3) "A is the same species as B" and "B is the same species as C" imply that "A is the same species as C." Such an equivalence relation is the only way to divide any whole population into distinct sets (which in our case are species). (1) and (2) are obviously true for our definition of species, but the reader should verify that (3) is true only if we eliminate asexuals and individuals living in very different places and times. Hence a proper partitioning of the individuals into species can be achieved only for individuals that share a common recent ancestry.

Knowledge of what species are is not equivalent to knowledge of how they develop. Does a large mutation suddenly change one individual sufficiently so that it is, by this one step, the first member of a new species? Or do we need an accumulation of very small steps? If we do need an accumulation of small steps, under what conditions does this take place?

Let us first consider the one-step formation of new species. Although many asexuals can benefit from polyploidy, which arises in one step so that markedly different forms appear at once and are perpetuated, the sexual animal species that we are considering must find this difficult, because at least two of these polyploids must be produced to breed with one another, which is very improbable. Plants, which frequently can self-fertilize, can and do make abundant use of polyploidy in the origin of new species, although the polyploid sets of species seem especially characteristic of places that have been rapidly opened for colonization, such as areas exposed by retreating glaciers. An interesting example is the formation of the marsh grass *Spartina townsendii* by fertilization of the British *S. maritima* by the American *S. alterniflora*. The zygote thus produced then doubles each chromosome, to form what is known as an allotetraploid. This allotetraploid is sufficiently different in habit from either parent so that it is rapidly taking over the English coastline. However, polyploidy seems of no significance to most animals in formation of new species and is by no means the only important method in plants. The role of polyploidy can, of course, be assessed by cytological studies.

No other type of "large" mutation, subject to phenotypic changes large enough to cause the origin of new species, has been witnessed,* and we next consider the accumulation of small changes.

Two populations can be distinct in several ways. They can be (*a*) actually separated so that during everyday existence their members

---

* The bizarre "homeotic" mutants known, for instance, to change *Drosophila* antennae to wings do not seem to produce the kind of difference seen between species. The claim that, in phylogeny, phyla arose first, then classes, and species last has been used as evidence for large "macromutations," but higher categories like phyla and class, do not seem to have any precise meaning, and the argument is a museum of confused semantics.

do not come across one another; (b) behaviorally isolated so that, although they meet one another, their types of reproductive behavior do not lead to successful breeding; (c) genetically or physiologically isolated so that no fertile hybrid offspring are possible. A notable achievement in the last thirty years has been the marshaling of nearly overwhelming evidence that for animals (a) precedes (b) and (c) in the normal species formation. Considered in terms of our understanding of what species are, this is not surprising, for we noticed that members of a species show a kind of cohesion due to genetic interchange. Some force seems to be needed to break this cohesion and, until it is broken by actually separating populations, genetic continuity prevents behavioral and physiological isolation. This is an argument to make the statement plausible. The empirical evidence has been gathered by Mayr (1963). It indicates, first that islands near the mainland often have two or more species resembling one mainland species. This is most easily explained in terms of repeated immigration by the mainland stock to the island on which populations are conveniently isolated from the mainland. The mainland, without this opportunity for isolation, has only one species. Again, evidence can be drawn from several pairs of populations, which are just on the borderline of being properly isolated species and which are most simply interpreted as species whose separation is not completed. For example, the myrtle warbler is an eastern American bird, and the Audubon's warbler is western. In the small area in which the populations meet there is some interbreeding, and the populations are clearly very closely related. If this really is a typical case of species whose formation is not yet quite complete, it is clear that geographic isolation has preceded behavioral and genetic isolation. Such closely related species nearly always differ in distribution. Finally, there are many species inhabiting each island of many-island archipelagos, whereas single isolated islands have only the single species descended from original colonizers. For example, ancestral finches landed apparently both on Cocos Island west of Central America and on the Galapagos island system, west of Ecuador, where opportunities for between-island isolation existed. On the Galapagos is the swarm of species of "Darwin's finches," which set Darwin thinking about evolution, whereas on Cocos only one species exists today.

We thus see populations becoming isolated, usually in different geographic areas, and while isolated diverging to such an extent that, after merging of the separate populations, they are free to go their own way instead of flowing together because of gene interchange. However, the story of what happens when these species meet is of wide interest and must wait until we have considered the interactions of populations.

Finally, we turn briefly to the question of the time it takes one species to split into two. When it takes place by polyploidy, as in the *Spartina* grasses already mentioned, it is completed literally overnight. However, many plants and most animals apparently form species by the longer route, for they make use of geographic isolation. How long does this take? We can proceed only a little way from known theory. Haldane's estimate (Chapter 2) of 300 generations per gene substitution gives us an order of magnitude, if we know by how many genes a pair of species differs. If they differ in 100 genes, at least $300 \times 100 = 30,000$ generations are required, but if they differ in 1000 genes, 300,000 generations are necessary. The average value may well be somewhere in between. We also saw in Chapter 2 that gene substitution can take place much faster than Haldane's estimate, especially in colonizing situations. Hence, occasionally, species can probably form much faster. The actual evidence is quite meager. The ages of various islands are known. The British Isles, for instance have been separated from the continent of Europe only since the last glaciers retreated. During this time Britain has not developed any certain new species, so that a period of 10,000 years or so seems inadequate. On the other hand, islands whose age is probably about a few million years seem to have many unique species. Hence somewhere between 10^4 and 10^6 years seem necessary. This seems perfectly consistent with Haldane's estimate, even if it is not very precise. The main point is that speciation clearly takes a long time, and critics of evolution have no right to demand that we exhibit species which have evolved within the memory of man.

THE NATURE OF ADAPTATION

Now that we have seen something of one of the consequences of natural selection (species) it is worth considering the other characteristics of organisms that natural selection has produced. In general, we call such phenotypic traits, which make their bearers fitter, *adaptations*.

Theoretically, if we believe that natural selection has shaped life as we know it, every aspect of biology could be treated as an example of adaptation. In fact, given great powers of deduction, a biologist by knowing only natural selection could predict quite closely the forms and function of each phenotypic trait. This clever biologist would never be absolutely correct, for four reasons: first, there may not have been enough time for the trait to be perfectly formed; second, some deaths, hence some of the evolutionary changes, occur to more or less randomly chosen individuals—this superimposes a certain unpredictable element onto the well-directed natural selection; third, mutations for some traits—such as wheels instead of feet—may not even theoretically exist; fourth, there is always a legacy of existing genes on which evolution must build. Still, we would like to be able to predict remaining properties of populations as outcomes of natural selection; that is, we would like to show that populations have adapted "best strategies," within the limits imposed by these constraints.

Actually, our knowledge is much too meager. Once we understand the workings of a trait, such as the eye or the numerical ratio of males/females in a population (to be considered later), we can marvel at the cleverness of these adaptations. But we can seldom predict what a vertebrate eye, for example, should look like, before we have ever seen one. That is, we often recognize a good adaptation when we see it, but we are seldom sufficiently good "design engineers" to say a priori what would be the best construction for a given trait. In this section we shall include a few examples on which we *can* make fairly accurate predictions and test them. Perhaps a twenty-first century population biology book will be able to include most of its subject under the heading of predictions from natural selection.

Before launching into the examples, it is worth defending the program of making predictions against claims that the unpredictable elements in evolutionary change are so overwhelmingly large that no interesting prediction can be made. The best evidence that prediction is possible comes from the marvelously close *convergence* which is so frequent between unrelated organisms adapted to similar ways of life. Figure 3-3 shows how closely the American meadowlark, *Sturnella magna*, and the African longclaw, *Macronyx croceus*, resemble each other, right down to the last detail of their form and coloration. Both are grassland foragers that prod the ground for food. Yet details of their internal anatomy make us believe that they are not closely related, and other members of the two families are very different. The precision and degree of convergence shows that, given a way of life, natural selection in separate situations produces almost identical results. This convergence would not exist if the unpredictable element were very large. Now, let us look at the examples.

A simplified version of the theory of sex ratio in vertebrates, first proposed by the British evolutionist Ronald Fisher, is a good initial example. As Fisher (1958) noted, each vertebrate has a parent of each

Fig. 3-3. Two cases of convergence. (a) Little auk, above, in wader-gull-auk order, closely resembles Magellan diving petrel, in albatross-petrel order. (b) Eastern meadowlark (*Sturnella magna*) (*above*) is an American Icterid; yellow-throated longclaw (*Macronix croceus*) an African Motacillid, entirely unrelated. (From J. Fisher and R. T. Peterson, *The World of Birds*, Doubleday, 1964, p. 47.)

sex, and each parent contributes equally* to the genetic makeup of the offspring. Thus males, as a group, make the same contribution as the females, as a group, to the ancestry of future generations. In other terms, the group of all males is "worth" the same amount as the group of all females. This means that if males should be temporarily only half as common as females, *each* male must be worth twice as much as each female. Thus, natural selection would favor an increase in proportion of males, until a 50:50 ratio were reestablished.

To make this clearer, we consider a population of several males but only two females, one of which has a gene that causes her to produce an excess of female children, and the other of which produces half male children and half female. To make matters precise, assume

---

* Except for the sex chromosomes.

female *A* produces 1 male and 9 females, whereas female *B* produces 5 males and 5 females. We shall now show that female *A* leaves fewer genes in the *grandchildren* than does female *B*, so that natural selection favors female *B*, with a 50:50 sex ratio. In the children's generation there is a total of 6 males and 14 females, and these 6 males must share equally with the 14 females in the production of the next generation. In other words, each grandchild of *A* or *B* must have one of the 6 males as a father and one of the 14 females as a mother. We pointed out that the grandchildren acquire about half their genes from male parents and half from female parents. Of the male half, *A* has contributed $\frac{1}{6}$ and *B* the remaining $\frac{5}{6}$; of the female half, *A* has contributed $\frac{9}{14}$ and *B* has contributed $\frac{5}{14}$. Thus the total contribution of *A* by way of both males and females is $\frac{1}{2}$ ($\frac{1}{6}$) + $\frac{1}{2}$ ($\frac{9}{14}$) = $\frac{17}{42}$, and the total contribution of *B* is $\frac{1}{2}$ ($\frac{5}{6}$) + $\frac{1}{2}$ ($\frac{5}{14}$) = $\frac{25}{42}$, which is greater than *A*'s contribution. In other words, if female *B* has a gene causing the 50:50 sex ratio of her offspring, that gene will be found in a greater proportion of grandchildren than will the allele possessed by *A*, and the 50:50 sex ratio will be favored. This argument, which can be modified to include sex-linked genes and differences in ease of rearing different sexes, holds for all sexually reproducing organisms and explains the generally equal proportions of males and females that we all observe. It also can be extended to explain the slight deviations from 50:50 sex ratio.

Another example that is not immediately obvious concerns the birthrates and death rates of organisms. Confronted with the fact that mammals such as elephants, lions, and whales have a low birthrate, but mice, rabbits, and the like may have dozens of young each year beginning when they are less than a year old, the man in the street* usually replies that elephants, lions, and whales lead a relatively safe life and do not need many offspring. It is more accurate, however, to ask which of two individuals is fitter: one which has many offspring or one which has few. Unless the many, by their very number, are correspondingly less likely to survive and reproduce, the parent of the many must be fitter. Thus, although many children are not needed, natural selection will favor as many as the parents can successfully rear. Why then, do elephants, lions, and whales have low reproductive rates? It must be because their large sizes and the demands of their long period of parental care make it impossible to have larger birthrates. Table 3-1 shows this for the European starling. Clearly the mother birds, on the average, lay that number of eggs that will result in the

---

* Even Darwin, the greatest of all population biologists, made this error.

maximum number of surviving offspring. This does not imply that clutch size will always increase under the action of selection, however, for a dense population may reduce its food supply, so that a smaller clutch leaves more survivors, each being better fed.

Table 3-1   Survival in Relation to Number of Young in Swiss Starlings
After Lack (1954)

Number of Young in Brood	Number of Young Marked	Number of Individuals Recaptured After 3 Months, per 100 Broods Ringed
1	65	—
2	328	3.7
3	1,278	6.1
4	3,956	8.3
5	6,175	10.4
6	3,156	10.1
7	651	} 10.2
8	120	
9, 10	28	—

We can pursue this argument a little farther and incorporate death rates and geographic distribution. We know how many eggs a bird will lay: as many as it can successfully rear. Where will this bird be able to live, then? It can persist wherever this birthrate exceeds the *inevitable* mortality due to hazards of weather, migration, etc. At each such place the population will then grow until the *additional* mortality accompanying the large population (perhaps disease, food shortage, gathering predators, or just a shortage of homesites) balances the excess of birthrate over death rate. We can make at least one simple prediction from this: where inevitable mortality is higher, which is often at high latitudes, birthrates will often also be higher. This is a very well-documented generalization (see Fig. 1-28 of Chapter 1). It explains why birthrate must be high at high latitudes, but not why it should be low near the equator. We must conclude that some other factor is responsible for the tropical species having small clutches (perhaps a less dense supply of food).

Finally, we can add migration to the picture. Why should a bird (or any other animal) migrate each year? Picture two individuals of the same species in the tropics in the early spring, one about to migrate north and the other to stay for the summer. Under what conditions will the migrant type be fitter and hence favored by natural selection? The migrant will be fitter if the benefits of the migration (mostly an untapped food supply) exceed the hazards of the migration. Obviously, just as many birds will migrate as will leave the food supply of the high latitudes more plentiful to a sufficient degree to compensate precisely for the hazards of the migration and of the more severe climate in higher latitudes. This is a prediction that we can make, but it has never been tested. It does lead to some understanding of the puzzle mentioned in the last paragraph, for now we see why there should be more food per bird in temperate regions.

We will end with a more traditional example of adaptation—the action of natural selection—mimicry. There are two extreme types of mimicry, named Batesian and Müllerian after H. W. Bates and F. Müller, and all kinds of intermediate types are possible. In Batesian mimicry, a "mimic" species deceives its predator by resembling a bad-tasting (or otherwise objectionable) "model" form which the predator has learned to avoid. Thus the mimic gains its advantage at the expense of both the predator and the bad-tasting model (which now suffers from attacks by predators who have experienced the palatable mimic). At the other extreme, Müllerian mimicry confers a three-way benefit. Here two bad-tasting forms by resembling each other are both fitter because the predator must learn to avoid only one form and hence is less destructive to both. There are some experimentally verified cases of mimicry. (That is, it has been shown that birds are deceived by the resemblance—on being trained on the palatable form, they readily try the unpalatable one and then spit it out; when trained on the unpalatable one, which they quickly learn to avoid, they will then avoid the palatable form). Besides these experiments, there is other striking evidence for mimicry: the resemblance is confined to external characters and to forms that occupy the same area. In spite of this evidence, many people do not believe mimicry exists—they claim that the resemblances are accidental or are caused by similar environments. (In part, their reluctance is brought about by the claim of uncritical enthusiasts that every resemblance of any kind is mimicry.) One worker, having tried Monarch butterflies himself and having found them palatable (although apparently not good enough to make them a part of his steady diet) claimed that this disproved mimicry involving this form as a model. Actually Brower (1958) had shown that birds quickly learn to avoid Monarchs and never actually eat them even when learning; because birds often eat butterflies, the mimicry is clearly adaptive. Various interesting predictions can be made, and tested, about Batesian mimicry. For example, mimics should be perfect in color and shade only if the predator has color vision. (Birds and some reptiles have color vision, but most mammals other than primates do not). Hence analysis of black and white pictures should reveal other cases of mimicry, which we with our color vision would not suspect. Furthermore, there might be mimicry of smell, sound, etc., which no one has looked for. We can predict that the effectiveness of the mimicry will decrease as the proportion of mimics increases, for then the predator will have relatively more experiences with the palatable forms. Brower has verified this (Fig. 3-4).

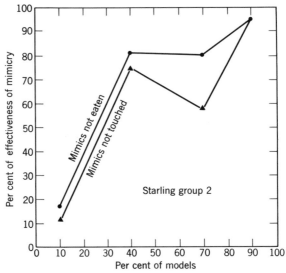

Fig. 3-4. Effectiveness of mimicry at different proportions of models and mimics. The per cent of effective mimicry is 100 times the ratio of the number of mimics not eaten (or touched) to the total number of mimics. (From Brower, 1960.)

Many of the patterns discussed in the remainder of this book will someday have to be explained as an outcome of natural selection—as a "best strategy" in a sense—but as yet we cannot do it.

PART THREE

THE FUNCTIONING
OF POPULATIONS

4

Population Integration

Cells and individual organisms obviously are highly organized units; they are bounded by skins and integrated by communication systems of nerves and chemical hormones. In contrast, populations have no tangible boundaries, and the communication systems are not so obvious. Some populations show almost no organization; annual plants or small crustacea swimming in the plankton appear to behave almost independently of one another. In contrast, the individuals in a colony of bees are completely dependent upon one another.

Let us begin by considering the communication between an individual and its mate and offspring. Sexual exchange of genetic information occurs in most, probably all, species sometime during their life cycle. It would be most efficient if individuals were able to recognize members of their own species and of the opposite sex. Much time, energy, and sperm would be wasted if cross-fertilization were completely random. In some species sperm is apparently scattered at random: grass pollen is dispersed by the wind, and sperm and eggs are entrusted to the whim of water currents in oysters and clams. Yet even these are not purely random events. Most of the pollen falls near the plant that produced it, and, because most plants live in aggregations, much of the pollen probably reaches another plant of the same species. Oysters do not release their eggs until sperm is detected in the current that is drawn into the respiratory passages.

The wastage is further reduced in most species by some kind of directed mating. A flower is a device that guides a pollinating insect from one individual plant to another of the same species. An animal is usually able to distinguish a member of its own species, its sex, and, in many instances, can recognize it as an individual.

However, recognition of a possible mate is not enough: mating requires contact between the egg and sperm, but most animals keep their distance from others of the same species. The reason probably is that they all require approximately the same kind of food and habitat, so that competition among them is strong. If we look at a row of birds perched on a telephone wire or a fence, it is obvious that they are spaced a small distance apart. This has been called the "individual distance" by Hediger (1955) (Fig. 4-1). It occurs in birds and mammals

Fig. 4-1.   Black-billed gulls (*Larus bulleri*) on the Taieri River, Otago, New Zealand. This photograph shows individual distances in gulls. It was taken during a breeding season; the gulls breed and feed just outside of this area. (Photograph by C. G. Beer.)

and also in some of the "lower" invertebrates, such as insects and crustacea. In animal species which have internal fertilization (most land animals), the "individual distance" must obviously be reduced to zero if mating is to occur. This is accomplished by communication with all manner of courtship signals. For example, the male jumping spider is much smaller than the female and would normally be eaten by her if he moved too close. By elaborate signaling from a safe distance, dancing or waving his limbs in a particular way, he establishes the fact that he is a mate, not food; she finally goes into a trancelike state, and he can safely approach her to mate (Fig. 4-2).

Once mating has occurred, there may be no further communication

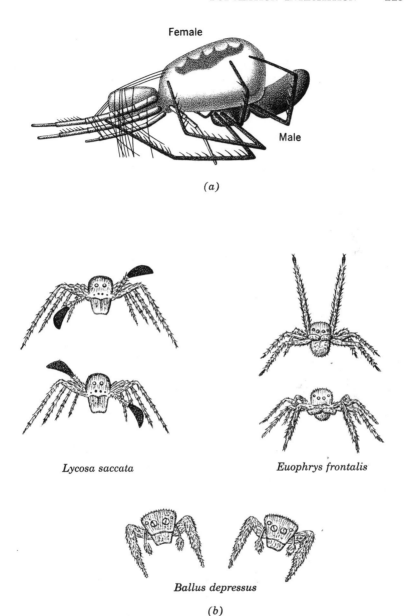

Female

Male

(a)

Lycosa saccata

Euophrys frontalis

Ballus depressus

(b)

Fig. 4-2. (a) The courtship of spiders. The male *Xysticus lanio* copulates with his female, after fastening her down with threads. (b) The courtship of spiders with good vision. The males display in this fashion before the female by sema- phore signals with the palps or forelegs, or by dancing to and fro. (From Marler, 1959.)

between the mates or with the offspring. Plants and most of the smaller animals are examples of this. However, in the social insects (bees, ants, and termites) and in the larger animals, there is a period of care of offspring. They may be merely fed and protected during their development and growth, ensuring a better survival during this vulnerable period. In addition, they may be given information that would increase their chances of survival under their own particular local conditions.

All this requires communication. The young must be able to signal that it is hungry, frightened, and so forth; the parent must be able to recognize its own young, locate it, and warn it of danger.

These cooperative activities of mating and care of young obviously increase the fitness of an individual by increasing the number and quality of its offspring. In most instances, however, fitness would not be increased by cooperation with individuals of other families. In fact, aggressive, antisocial interactions between families is the usual pattern. The "individual distance" is enlarged, and mate and offspring are usually tolerated within it. It is then called a "territory." This territory is defended against encroachment by individuals outside the family. It may be large enough to include all the food necessities of the family; eagles defend huge territories in which they do all their feeding. In other species, which feed in places where nests cannot be built, it may be very small. Sea birds defend only a small zone around the nest, and seals only a stretch of shoreline in which the male, his mates, and offspring can rest.

The defense of a territory means that the resource which is in short supply, whether it be food or space, is parceled out to a select few. These then have enough resources to maintain themselves and to rear young; the rest of the population does not breed and may starve or may be exposed to predators in unfavorable habitats. The alternative to this behavior is a "scramble" for the resources, with the possibility that no individual will obtain enough to survive.

In contrast to these populations, there are others in which cooperation between individuals of different families exists. For such cooperation to have evolved under natural selection *more offspring must have been left by those individuals which associated in groups larger than their own families.* Social behavior must have conferred advantages on individuals.

There are many examples of such advantages. Small animals may increase their "effective size" by operating in bands. Small birds in a group drive away larger hawks and owls. Predators, such as wild hunting dogs and wolves, can run down and kill much larger animals by operating in packs.

Larger effective size may render animals more independent of the physical environment. Beehives and termite mounds are kept at almost constant temperatures by the activities of large numbers of small animals. Birds and mammals keep warm at night by huddling in groups. Toxic substances in water were removed by groups of goldfish, whereas single fish succumbed (Allee, 1951).

A group is not only larger than a single individual, but has more sense organs dispersed over a larger area. Whereas a solitary individual must continually interrupt its feeding in order to guard against approaching predators, those in a group can spend more time feeding and resting, because only a few need to be on guard at any one time.

When food appears in large amounts at irregular intervals, a widely scattered set of sense organs and a means of communication is an advantage. A large carcass may be too much for one vulture to eat; when one vulture discovers and approaches it, others, which are scattered hunting over a wide area, are attracted by its activities and join the feast. The discovery by a honey bee of a new bloom of flowers is communicated to the other workers in the hive, and the nectar and pollen are quickly gathered.

Where predators are separated from their prey at intervals, the grouping of the prey may decrease the probability of the predator finding them again. For example, many marine organisms occur in groups: fish in schools, snails and burrowing crabs in "herds" on beaches. During low tide or at night, predators may become separated from their prey. When a predator can again search for its prey, the chances of its finding them are probably less if they are grouped than if they are uniformly scattered. It is obvious that advantages often accrue to an individual which cooperates with others in a group composed of several families.

If the cooperative behavior among families is genetically determined, so that by operating in such a group an individual leaves more offspring, social behavior will evolve by natural selection. This seems to be the simplest explanation of the idea of "group selection" that we discussed earlier.

Once a gregarious, communicative life is established, the animals reap another benefit: they learn from one another. Living in groups, they observe others' activities more closely than if they lived singly. By so doing they are evidently able to profit from the experience of others, as shown in the following study by Klopfer (1962).

Greenfinches and great tits are common birds in England; the greenfinch is solitary in its feeding, and the great tit is gregarious. Klopfer first showed that an isolated individual of either species could rapidly learn to avoid seeds, which he had filled with aspirin to make them distasteful, when they were placed on a colored dish. That is, the birds soon avoided all seeds on the colored dish, even though they could not distinguish the distasteful ones. The normal seeds were in a plain dish, and the birds continued to eat these. Next, he repeated his experiments with new birds, this time by putting two individuals into adjacent compartments of a screened cage in full view of one another's learning. The results were remarkable: the normally gregarious tits learned even faster than when isolated, each apparently able to benefit from witnessing the other's mistakes. The normally solitary greenfinches became hopelessly confused and never learned to avoid the distasteful seeds. Apparently, when one witnessed the other trying the distasteful seeds on the colored dish, he ignored his own previous experience and tried them himself. This, in turn, confused the other, and so forth. This may help to explain the rapid spread of the habit of great tits (and other tit species, too) of stealing cream from milk bottles left on window sills. The birds fly to the bottles and pull off or tear the cap and sip the cream from the tops. They are so effective at this that, if the bottles are left unprotected, an inch or two of cream is removed within a few minutes. This habit has spread over most of England within this century; the rapid spread is probably due to the tit's ability to benefit from observing the success of others in finding food. Notice that the birds need not copy the feeding behavior—tearing milk bottle caps is similar to their normal behavior of tearing bark off trees to find the grubs underneath —and all that needed to be passed from bird to bird was the secret that milk bottles are a good feeding place. Klopfer's experiments show that only a social bird like a tit could benefit from this kind of experience. This gives a clue to the kind of benefits that soon follow from a gregarious life.

This keen observation and good judgment of others' behavior is the really important by-product of social life. It leads to the ability to discriminate among various signals of other individuals, which eventually leads to elaborate communication systems. These, in turn, make sociability even more beneficial, so that once the trend to social behavior is started it gains momentum. Finally, the resulting communication is a cause of the reduced confusion in social populations. All this, of course, requires a complicated nervous system and is, therefore, evident especially in vertebrates, mollusks, and arthropods.

However, these advantages of social life must be balanced against the danger that the available food or space may be spread amongst so many individuals that none have enough. This is solved by territoriality in some species. Nevertheless, it may be impossible to set up territories if the group must continually be on the move in their search for food, as many grazing mammals or schools of fish are. In these groups a rank order of privilege or "peck order" is developed within the group. When food is found, the high-ranking individuals have first pickings; individuals of lower rank which challenge the dominant ones are attacked or threatened and driven off. Chickens, pigeons, fish, cows, wolves, and monkeys all show this type of organization. The dominant individual, shows its social position by the way it struts, holds its head or tail, etc.

If a group is newly formed, for example, when hens are first placed together in a yard, there is much fighting. Then, as a dominance hierarchy is established, the fights are reduced to symbolic threats and submissions. This is obviously a more efficient organization, because less energy is wasted in conflicts and less damage is done. Once the peck order is established, chickens eat more, gain more weight, and lay more eggs than during the initial phases (Guhl, 1953).

This process of population integration has reached its culmination in man, where language has allowed a great body of traditional information to be transmitted. Etkin (1964, page 290) has suggested that the division of labor between the sexes in hunting animals such as wolves, where females spend much of their time nursing the young and males do most of the hunting, may be conducive to the development of language. Events during the hunt may need to be communicated to those back home who had not witnessed them and vice versa. He and others have suggested that if early man had shifted from gathering food, the common way of life of most subhuman primates, to hunting it, the development of abstract language would have been encouraged. The communication of the direction, distance, and kind of food by the scouts to other worker bees, although not an interfamily example, illustrates this point.

It should be evident by now that, like cells and organisms, populations are kept together and integrated by systems of communication. The degree of organization varies widely, but it seems obvious that complete independence of individuals is inefficient, at least in a sexually reproducing species.

5

Population Growth and Regulation

INTRODUCTION

It is a matter of common observation that the numbers of any species change in time. Sometimes the change is an orderly growth and decline: changes in the number of houseflies from spring through autumn, for example. Sometimes the changes are violent and erratic: the arrival of a destructive horde of grasshoppers in a field, or a mouse plague. Houseflies increase in late spring because the birthrate exceeds the death rate; the reverse happens in the autumn. At the same time robins increase and decrease for these reasons and also because they migrate in and out of an area. Sudden changes are due mainly to mass movements, as happened when gulls arrived and destroyed a plague of grasshoppers during the early settlement of Utah. We shall consider the causes of population change one at a time.

BIRTH

When there are no movements in or out of the area and no mortality, change is a function of birthrate alone. For example, when a single bacterium is placed in a flask of suitable culture medium, it in time divides, producing two daughter cells. Each cell grows to the size of its parent, then divides again, and so on in the succession 1, 2, 4, 8, 16 . . . ; doubling occurs in each generation—about every twenty minutes

120

in the usual culture conditions. Thus the population N at the end of T generations will be 2^T, which can be written

$$N_T = 2^T = e^{\log_e 2^T} = e^{T \log_e 2} \tag{1}$$

where e is a number equal to $2.71828 \ldots$, the base of the natural logarithms.

When this kind of population growth can be regarded as continuous, the rate of growth at any instant is expressed as

$$\frac{dN}{dt} = rN \tag{2}$$

Where N is the number present, t is time, and r is the instantaneous rate of increase. Thus r is an interest rate, and is the "fitness" described in Chapter 2. The expression on the left side of the equation is calculus shorthand for "the rate of change in N with respect to t." The right side of the equation indicates that the rate of change is a function of the population size, N at time t. The rate of increase, r, is equal to the rates of (birth plus immigration) minus (death plus emigration); in this section we disregard the last three of these or at least assume their cumulative effect to be constant. To calculate the size of the population at any time, we can use formulas from calculus to integrate this equation, with the following result

$$N_t = N_0 e^{rt} \tag{3}$$

where N_0 = number at time zero, N_t = number at time t. Clearly, this is the same as Equation 1 for the bacterial case, where $N_0 = 1$ and the time unit is one generation to that $T = t$. Then $r = \log_e 2 = 0.693$.

Expressed in logarithmic form, Equation 3 becomes

$$\log_e N_t = \log_e N_0 + rt \tag{4}$$

which is recognizable as the equation of a straight line of slope r, when the logarithm of the population size is graphed against time, as in Fig. 5-1. This growth is termed "logarithmic" or, better, "exponential."

When, as in this example, r is positive, and the conditions remain the same, on exponential growth the population would keep increasing ever faster. Calculations of the consequences of such growth make amusing exercises. For example, if bacteria reproduce every 20 minutes, exponential growth for 36 hours would produce a layer of bacteria a foot deep over the entire earth, which within the next hour would be over our heads; ultimately, given up to a few thousand years, any species of plant or animal would weigh as much as the visible universe and would be expanding outward at the speed of light! Charles Darwin calculated

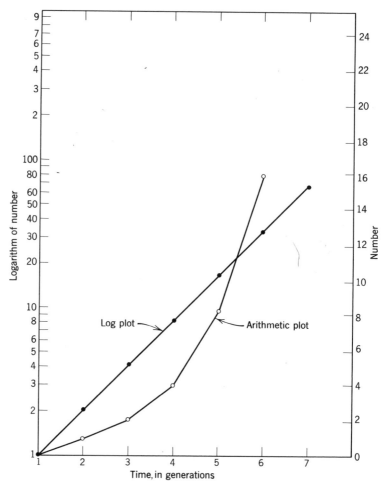

Fig. 5-1   Exponential population growth, plotted in two ways.

that if one pair of elephants started breeding at age 30 and survived until 100, meanwhile having produced on the average six young per head, there would be nearly 19 million elephants alive after 750 years.

Because we do not observe such numbers, it is obvious that continued exponential growth of populations does not occur. It may be approached, however, when a species is newly introduced into a region where conditions are favorable. Bacteria newly introduced into a favorable culture medium grow exponentially awhile. Thus 435 striped bass from the Atlantic Ocean were planted in San Francisco Bay in 1879 and 1881; in 1899 the commercial net catch alone was 1,234,000 pounds (Merriam, 1941). Clearly, their growth was nearly unimpeded.

Such "population explosions" may occur for a short time whenever conditions change for the better. The floating unicellular plants of some temperate oceans show a "spring bloom" of population growth each year. The same rapid growth occurs in the spring in ponds or on land; it is most striking in high latitudes where the seasonal change is greatest. This exponential growth never lasts very long before it is slowed down by a shortage of some essential requisite, by the attacks of a predator, or by some other limiting factor.

In these examples from natural populations both birth and death were obviously occurring. Before we consider the interaction of birth and mortality, we shall discuss mortality alone.

MORTALITY

In some natural situations, such as those at high latitudes, nonmigratory species may produce all their young in a short period in the spring. The subsequent decline in numbers can then be studied without the complications of birth and migration. If the logarithm of the numbers is

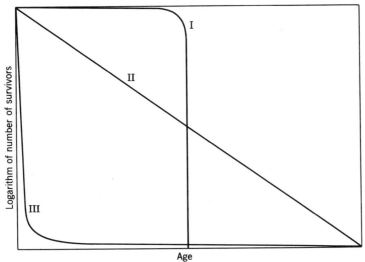

Fig. 5-2. Schematic representation of theoretical types of survivorship curve. (After Deevey, 1947.) The survival axis can be graduated either arithmetically or logarithmically, but the logarithmic scale is more instructive, in that a straight line implies equal rates of mortality with respect to age.

plotted against time (or age) a survivorship curve is produced. Its shape may vary from the straight line of exponential decrease found in bacterial populations and, surprisingly, in many birds, to the convex or concave shapes shown in Fig. 5-2. The straight-line survivorship curve indicates constant mortality, independent of age. The larger vertebrate animals, including man, may show the convex pattern, in which little mortality occurs in the early portion of the life-span. Smaller organisms generally have a very high mortality rate in the early ages, so that only a few reach old age; their curve is concave (Deevey, 1947). If detailed records are possible, as in the barnacles studied by Connell (1961a), "sinuosities" may occur, corresponding to changes in the causes of mortality at different ages (Fig. 5-3).

Mortality records can be shown in another form called, paradoxically, a "life table." This shows the mortality rates, survivorship, and expectation of further life for a population; an example is given in Table 5-1. Life tables were developed by the life insurance industry for calculating life expectancies, and they have only recently been applied to organisms other than man; several have been constructed for animals, but there is as yet no life table for any plant, although foresters have assembled some of the data.

If the survivorship curve is a straight line, the mortality rate does not vary with age. This probably never happens in nature; thus the curves for birds did not take into account the higher mortality of the young. As the mortality rate varies with age, so also does the birthrate. These variations make it necessary to know something about the age structure of a population.

AGE STRUCTURE

If the life-span of a species is greater than its period of reproductive maturation, the generations overlap. In these instances the population has an "age structure," which is described by the proportion of the total population in each age group.

Each group has a specific birth and death rate, which may change with age. The birthrate in humans is highest around 20 years of age and lower before and after; death rates are lowest around 12 years and highest in the first year and in old age. These age-specific rates probably remain fairly constant, if environmental conditions remain the same. If conditions change, a new schedule of age-specific birth and death rates is established. For example, with the industrial revolution and the improvements in public health, human death rates decreased,

Table 5-1 Life Table for a Typical Population of Balanus balanoides, Based on the Observed Survival of Adult Barnacles Settling on a Cleaned Rock Surface in the Spring of 1930.*

x	x′	d_x	l_x	1000 q_x	e_x
† Age (Months)	† Age as per cent Deviation from Mean Length of Life	Number Dying in Age Interval out of 1000 Attaching	Number Surviving to Beginning of Age Interval out of 1000 Attaching	Mortality Rate per Thousand Alive at Beginning of Age Interval	Expectation of Further Life (Months)
0–2	−100	90	1000	90	12.1
2–4	−83.5	100	910	110	11.3
4–6	−67.0	50	810	62	10.5
6–8	−50.4	60	760	79	9.1
8–10	−33.9	80	700	114	7.8
10–12	−17.4	160	620	258	6.7
12–14	−0.9	80	460	174	6.7
14–16	+16.0	100	380	263	5.9
16–18	+32.2	50	280	179	5.7
18–20	+49.0	40	230	174	4.7
20–22	+65.4	100	190	526	2.4
22–24	+82.0	60	90	667	1.9
24–26	+98.8	20	30	667	1.8
26–28	+115.0	8	10	800	1.4
28–30	+132.0	2	2	1000	1.0

† Survivorship data given graphically by Hatton were smoothed by eye, and values at every other month were then read from the curve. The original observations were made at irregular intevals during three years.

* The population is that at Cite (St. Malo, France), a moderately sheltered location, at Level III, at half-tide level. The initial settling density (2200 per 100 cm²) is taken as the maximum density attained on May 15. Mean length of life 12.1 months. Data from Hatton (1938), analysis by Deevey (1947).

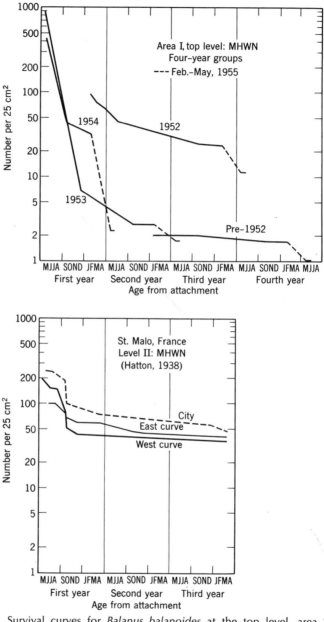

Fig. 5-3. Survival curves for *Balanus balanoides* at the top level, area 1 (where it was mixed with *Chthamalus stellatus*); curves from a similar level in Brittany were replotted from Hatton (1938). (After Connell, 1961a.)

as did the birthrates. The same probably happened when humans changed from a hunting and gathering economy to an agricultural one. As a result, the human population increased in these two periods, as shown in Fig. 5-4.

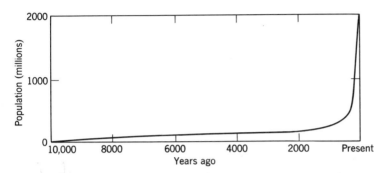

Fig. 5-4. Arithmetic population curve plots the growth of human population from 10,000 years ago to the present. Such a curve suggests that the population figure remained close to the base line for an indefinite period from the remote past to about 500 years ago, and that it has surged abruptly during the last 500 years as a result of the scientific-industrial revolution. (Deevey, 1960.)

Over spans of time of the order of hundreds of years, which may correspond to the maximum length of a generation in long-lived plants, it is probably safe to say that environmental conditions will not change drastically, except as a result of man's activities. In places where there has been little disturbance by man, therefore, it is probable that the age-specific rates of birth and deaths have also remained constant, on the average. In these instances, a particular age structure will be reached and maintained. The most easily understood situation is one in which the population size has remained constant, the number of births equaling the number of deaths. Here a "stationary age structure" develops, and each female replaces herself (and her mate) exactly once in her lifetime. In any natural population it is unlikely that the replacement rate will be exactly 1.0 in any particular generation, but over the long run it must be so, because otherwise the population would either decrease to extinction or continue indefinitely to rise.

If the age structure appears to be remaining constant from year to year and birth and death rates are known, a calculation of the replacement rate can be made. This is the expected number of female offspring left by a female during the course of her life. A column of age-specific birth rates, m_x, is added to the life table. This is the number of female offspring born to a female during the interval of age from x to $x + 1$.

Setting the initial survivorship, l_0, equal to 1.0, the formula for the replacement rate, or net reproductive rate, R_0, is

$$R_0 = \sum_{x=0}^{\infty} l_x m_x \qquad (5)$$

Table 5-2 shows some calculations of R_0 for sowbug and salamander populations. The age structure of the sowbug population had remained

Table 5-2   Calculations of the Net Reproductive
Rate, R_0, in Natural Populations

a. Five species of the salamander genus Desmognathus
(Organ, 1961)

Age (Years)	l_x	m_x	$l_x m_x$	$R_0 = (\Sigma\, l_x m_x)$
D. quadrimaculatus				
0	1.0	0	0	
5	0.053	15.5	0.851	
7	0.0088	15.5	0.136	
9	0.0013	15.5	0.021	1.01
D. monticola				
5	0.049	13.5	0.661	
7	0.0075	13.5	0.101	
9	0.0012	13.5	0.016	0.78
D. fuscus				
5	0.034	11.5	0.391	
7	0.0024	11.5	0.028	
9	0.0001	11.5	0.001	0.42
D. ochrophaeus				
5	0.120	5	0.600	
7	0.027	5	0.135	
9	0.0046	5	0.023	0.76
D. wrighti				
5	0.200	3	0.600	
7	0.035	3	0.105	
9	0.001	3	0.004	0.71
b. Armadillidium vulgare (Paris and Pitelka, 1962)				
0	1.0	0	0	
1	0.0555	6.25	0.347	
2	0.0058	85.05	0.493	
3	0.0008	201.96	0.161	
4	0.0001	237.50	0.024	1.02

relatively constant over three years, so that it seemed reasonable to assume a stationary age distribution; R_0 was approximately 1.0. In the five species of salamanders the more terrestrial species were smaller, produced fewer eggs, and had lower mortality rates than those living in or near the streams. In four of the five, the author considered that R_0 was about 1.0, taking into account the errors due to sampling. Thus the expectation of a balance between birth and death rates appears to be confirmed in natural populations.

Under any schedule of constant birth and death rates a particular age structure will develop and, surprisingly enough, the population will come to grow smoothly and according to Equation 2, where r has to be calculated from the schedule of births and deaths by a rather complicated formula. In this ideal optimum situation, in which a population is growing exponentially without limits, the r in Equation 2 is called the "intrinsic rate of natural increase." The age structure which develops then is called the "stable" age distribution. Some populations of insects and rodents have been reared in the laboratory under what appeared to be the best conditions. The maximum r measured from these populations was probably a good approximation of their intrinsic rate of natural increase. In general, this rate is greater as the body size decreases, being about 50 per individual per day in bacteria and about 0.01 in rats.

AGE AND REPRODUCTION

When most of the young are born during one fairly short period of the female's life, we have a reasonably accurate, simple way of calculating r (of Equation 2) from the schedule of births and deaths. We defined R_0 to be the expected number of female offspring left during her life by a newborn female and pointed out that it can be calculated by Equation 5.

In each generation the population increases to R_0 times the existing size, N_0, as in the equation $N_1 = N_0 R_0$. In T generations, $N_T = N_0 (R_0)^T$. Suppose now that these R_0 female offspring are born to mothers of age A. Then in an interval of t years there will be t/A generations, during each of which the population increases R_0-fold. Hence the population increases according to the equation (where N_0 is the initial population size)

$$N_t = N_0 (R_0)^{t/A} = N_0 e^{(t/A) \log_e R_0} = N_0 e^{(1/A)(\log_e R_0)t} \qquad (6)$$

so that r (of Equations 2 or 3) is given by $r = (1/A) \log_e R_0$.

If the population remains constant in size, $R_0 = 1$, so that $\log_e R_0 = 0$ and, no matter what the value of A, $r = 0$, which we already knew to be true. When R_0 is near to 1, $\log_e R_0$ is near to zero, and a moderate change in age at reproduction, A, cannot change r by a very great amount. If, however, R_0 is rather large, a slight reduction in the age at reproduction will produce a profound increase in r.

We shall illustrate this with a human example. In a growing population, a woman who has her first child when she is 18, and has another each year until she has five children, contributes as much to the rate of population growth as another woman who has her first child when she is 30 and has ten children in ten years. Because, by Equation 6, for the first woman A is about 20 (for the middle child) and $R_0 = 2.5$ (the number of female children); thus $r = \frac{1}{20} \log_e 2.5 = 4.57$. Similarly, the second woman has $r = \frac{1}{35} \log_e 5 = 4.53$. The age at reproduction is only important in a growing population, as we showed earlier. If the population size were constant, it would not matter at what age the mother had her children. True population control can only be achieved by strict limitation of family size.

Human females stop reproducing long before they die. This is not so in other animals and plants. (For example, female rhesus monkeys continue bearing young regularly into physiological old age.) The reason probably is that any small deleterious mutations, whose effects do not show until after an individual has stopped reproducing, cannot affect its fitness and thus will not be eliminated by natural selection. These will accumulate and will tend to cause rapid aging once an individual stops reproducing. We might speculate that prolongation of life beyond the reproductive age occurs only when the postreproductive individual can increase the fitness of his offspring by the *nongenetic* communication of information. In the maintenance of the complex social organization of humans, the accumulated fund of learning and experience of old people may be valuable enough for this to have happened.

In natural populations, r fluctuates around zero. If r were continuously positive or negative for a protracted period, the population would either become dangerously high, exhaust its resources, and "crash," or else it would decrease to extinction. Therefore, natural selection must act to prevent protracted positive or negative values of r.

IMMIGRATION AND EMIGRATION

Up to this point we have dealt with populations as if each were isolated from all others of the same species. A population of lizards on an isolated island might be in this situation, but even they must have moved there at least once. In many situations birthrate is supplemented or even replaced by immigration, and death by emigration. Populations of sessile or sedentary organisms without vegetative reproduction, such as annual plants on land and most of the plants and animals living on or in the bottom of lakes and the sea, increase by immigration. They release gametes or young stages into the air or water where they are carried about (sometimes on or in other animals) and eventually settle down. This sounds rather haphazard, and there is undoubtedly much waste of seed and young. Therefore, it is not surprising to find that some of these animals are able to choose the right habitat.

Where changes in climate are so severe as to render continual occupation impossible, some species escape during the harsh intervals and return when conditions improve. In polar regions or at high altitudes migratory populations of birds, butterflies, large mammals, etc., grow and decline in this way, supplemented, of course, by birth and death. This means that there will also be changes in populations at low altitudes and latitudes when these animals arrive.

BODY SIZE, REPRODUCTION, AND MORTALITY

There are obvious advantages to being large, and the general trend of evolution, shown by most fossils, is toward larger size. However, the existence of a multitude of small organisms shows that these advantages are not all one-sided. Food often occurs in scattered small bits, which a small organism can eat more efficiently. Only when tiny bits of food are aggregated in clumps can larger animals, such as anteaters or whales, live on them. Some places, such as the soil or the insides of other animals, are habitable mainly by small animals. Small annual plants in deserts can grow and produce seeds in the short favorable period after a rain; for the rest of the year most of the space between the scattered bushes contains only their drought-resistant seeds.

However, small organisms are more at the mercy of the environment than large ones. Small animals can be eaten by any bigger predator; they can be pushed around by currents and winds and trapped by surface tension. With a larger surface-volume ratio, they are more affected by changes in salinity, moisture, and temperature and have higher metabolic rates. All these risks mean that smaller organisms have shorter life-spans, so that they usually have the chance to reproduce only once in their lifetime. Because their life is short and uncertain, most of their energy is channeled into a single reproduction.

Large organisms, on the other hand, channel a greater proportion of their energy into structures and activities that make them relatively independent of changes in the environment. The energy expended on acquiring protective bark, spines, or fur, on osmotic work, the maintenance of body temperatures, and parental care and teaching during the slow development of offspring, leaves less energy for reproduction. The price paid for longer life is a lower reproductive rate. Most large organisms reproduce several times during their lifetime; this is an obvious necessity in species such as elephants or the California condor, which produce only one young at each birth.

Some small organisms reproduce more than once in their lifetime. The heavily armored barnacles and molluscs of temperate intertidal shores may live for several years, reproducing each year. Internal parasites, which are protected inside their host, may reproduce for a long time; tapeworms probably live for several years, producing millions of eggs.

In contrast, a few large organisms reproduce only once. The Pacific salmon, the Atlantic eel, and the sea lamprey almost always die after a single spawning. Adults of these fish expend a great amount of their energy in a long reproductive migration, the salmon and the lamprey from the sea into fresh water and the eel from fresh water to the sea. The evolutionary strategy which these species employ resembles that of the annual plants: relatively more energy is channeled into reproductive activities and less into maintenance.

REGULATION OF THE SIZE OF POPULATIONS

Some species are common, others are rare; each has a characteristic level of abundance in a region. However, the numbers of any species do not remain constant at this level, but fluctuate to a greater or lesser degree. Figures 5-5 to 5-7 show some long-term records of the size of populations.

One of the central questions of population biology is: what determines the size of a population? The answer is: different causes under different circumstances. At one extreme there are organisms living under very harsh conditions, such as small insects in arid conditions; their populations grow during the short wet season, until the drought kills the plants upon which they feed, and the insects disappear. They are completely at the mercy of their physical environment, and the size of their population at the end of the wet season depends on their intrinsic rate of increase (the r of Equation 2) and the length of the growing season. Figure 5-6 shows the changes in

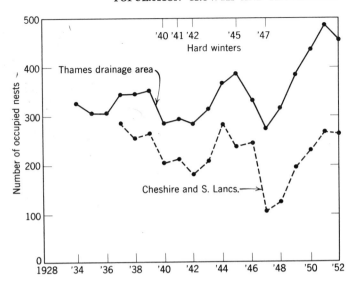

Fig. 5-5. Breeding population of the Heron (*Ardea cinerea*) in England (data from British Trust for Ornithology and W. B. Alexander, reanalyzed). (Lack, 1954.)

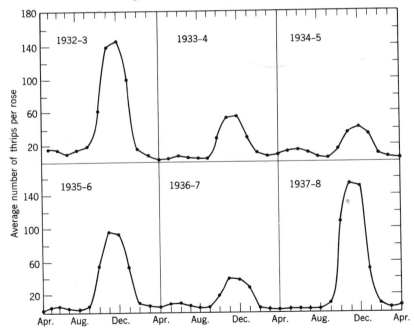

Fig. 5-6. Seasonal changes in a population of adult thrips living on roses. (Andrewartha and Birch, 1954.)

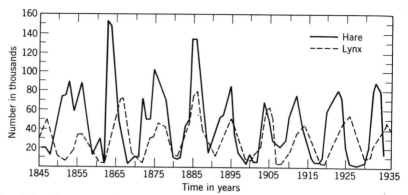

Fig. 5-7. Changes in the abundance of the lynx and the snowshoe hare, as indicated by the number of pelts received by the Hudson Bay Company. This is the classic case of cyclic oscillation in population density. (After Odum, 1959.)

numbers of thrips on roses in South Australia. A new colony is established by animals dispersing from other colonies nearby. If we know the local weather pattern, the distance to an existing population, the rate of mobility, and the intrinsic rate of increase, we can make predictions about the establishment and size of particular local populations. It is difficult, however, to make generalization beyond the local scene. The book of Andrewartha and Birch (1954) deals mainly with these kinds of populations and gives many examples.

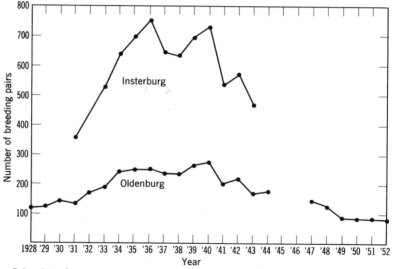

Fig. 5-8. Breeding population of the White Stork (*Ciconia ciconia*) in parts of Germany. (Lack, 1954.)

In contrast to this situation are those where organisms are not at the mercy of their physical environment. Large organisms in fairly harsh environments and most species in conditions such as the warm, wet, more evenly favorable tropical regions are relatively independent of their physical environment. The challenges that polar bears face in the arctic regions are starvation and predation by hunters, not extreme cold. Small fish living in the extremely stable environment of tropical coral reefs face similar challenges.

Under these circumstances the size of a population is determined by interactions either among its members or with other species of competitors, predators, or parasites. The simplest scheme is a self-regulating population; when the numbers are high the population decreases and when the numbers are low the population increases. We can illustrate this process by a set of diagrams modified from Ricker (1954). To use them, we must assume that the population density is the sole cause of population change and that the changes occur in annual jumps. If the population size, N, of a species in a given area is plotted on the abscissa, and the change in population, ΔN, which occurs when the population reaches size N, on the ordinate, a graph of the kind shown in Fig. 5-9 must result. The point E represents an equilibrium population density—one causing no change in population. From this graph alone we can plot the sequence of population sizes of a population. This is done in Figs. 5-10 and 5-11.

Beginning with point A, the population increases by the amount of the length of the vertical dotted line above A. B is precisely the same distance to the right of A. From B the population increases more, as the dotted line shows, and C is this new distance beyond B. From C, the population increases only slightly, which happens to bring us to the point E from which the population never strays. If, instead, the

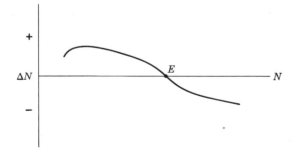

Fig. 5-9. The one-generation change, ΔN, in numbers of individuals is plotted against the number, N, of individuals preceding the change. When $N = E$, there is no change. The curve is hypothetical.

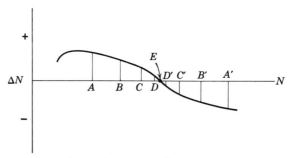

Fig. 5-10. A population of initial size *A*, will successively be of sizes *B*, *C*, *D*, and, perhaps, *E*. A population of initial size *A'* will successively decrease to *B'*, *C'*, *D'*. Notice that point *B* is placed so that the distance from *A* to *B* equals the height of the line segment above *A*. *C* is calculated from *B* in the same way and so on.

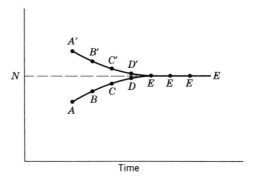

Fig. 5-11.   The results of Fig. 5-10, plotted with time as a coordinate.

population had initially been *A'*, it would have decreased to *B'* to *C'*, and, thence to *E*.

This graph, then, represents a population that approaches *E* from either side. Figures 5-12 and 5-13 are the corresponding figures for a population which changes more violently. Here, with the graph intersecting the X-axis more steeply, the increase when the population size is *A* is sufficient to exceed *E*. From this point *A*, it rises to *B* and then drops to *C*, etc., as shown in Fig. 5-14. Clearly, these oscillations about *E* will occur whenever the graph intersects the X-axis with a slope steeper than 45° (i.e. the tangent of the angle must be greater than 1), for then the vertical dotted lines will have greater length than the distance to *E*. In fact, if the lengths of these vertical lines are more than twice the distance to *E*, the oscillations will become more violent. The condition for this then is that the tangent of the angle of intersec-

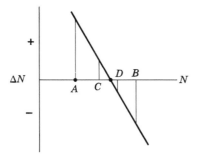

Fig. 5-12. This figure differs from Fig. 5-10 only in that the line is steeper. Successive populations A, B, C, D now oscillate. The oscillations are basically due to the time lag in the population response. A population which grows continuously, instead of in discrete generations, will not oscillate in this way.

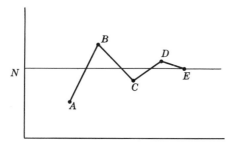

Fig. 5-13. This is the time curve plotted from Fig. 5-12.

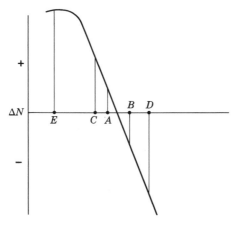

Fig. 5-14. This curve differs from Fig. 5-12 in being still steeper. The corresponding time curve is on page 138.

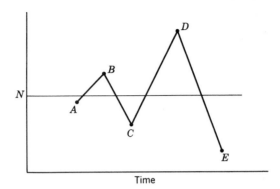

Fig. 5-15.   Calculated from Fig. 5-14, showing increasing oscillations.

tion should be greater than 2, or, from trigonometric tables, the angle of the intersection should be greater than 63° (Fig. 5-15). Obviously, bends in the graph influence this last conclusion; the reader can try various shapes of graph. In fact, we can here measure the stability by the amount by which the angle of intersection deviates from 45°.

An example of this is provided by the ovenbird population changes shown in Fig. 5-16, based on censuses over a period of 15 years in Ohio. Because this is a real environment, it is not constant; hence the points only cluster about the "true" graph. However, it does indicate clearly how populations can conform to the theory.

This scheme assumes that change is influenced by population size alone and that it occurs in jumps. In continuous, rather than jumplike, growth, no oscillations would have been produced. One mechanism by which the changes might be produced in animals has been proposed by Christian (1959). In denser populations, animals meet more often. If such meetings caused glandular malfunctions, which reduced birth rate or increased death rate, the population would decrease when it was dense and increase when it was sparse. Small rodents fight when they meet, and laboratory experiments have shown that the condition of the adrenals and other organs is changed as a result of the fights, with consequent damage to offspring.

Another possible factor regulating the size of those populations that are not completely at the mercy of their physical environment is the supply of some requisite, such a food, an essential mineral, or a growth factor. The explosive increase in numbers of plants in the plankton following addition of mineral nutrients and the subsequent increase in numbers of animals feeding on them indicate that populations are regu-

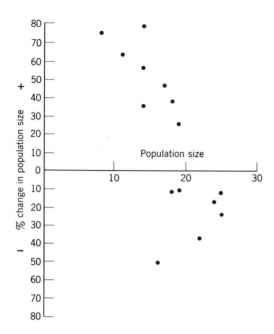

Fig. 5-16. Increases and decreases over an 18-year period in the ovenbird population of a wood near Cleveland, Ohio, are plotted against the population size preceding the increase or decrease.

lated by the amount of some limited resource. For example, one study (Pitelka et al., 1955) showed that large predatory birds in the arctic regions were able to breed when the population of their prey, lemmings, was high but not when it was low.

Although the supply of food is undoubtedly one of the main determinants of the size of a population, we rarely find starving animals in natural conditions. This in itself does not mean very much, because we rarely find dead or dying animals either. Death and decomposition are rapid processes under most conditions. However, the usual situation is that most organisms are active and apparently healthy in the wild. For this reason Wynne-Edwards (1962) has suggested that most animals somehow keep their population below the level at which they would begin to starve. He thinks that they regulate their numbers so as to avoid overeating and destroying their food supply. This may be accomplished in two ways. Territorial behavior and/or peck orders serve to parcel out the available food to the stronger members, so that they have enough to survive and breed, whereas the rest are excluded from breeding. Thus the production of offspring is adjusted to the food supply. Red grouse set up smaller territories and thus more bred in a heather moor in Scotland in years when the food supply was good than when the food was lean. There is also good evidence that there are more birds capable of breeding than actually do so. Presumably, these are extra birds that could not secure a territory. This mechanism acts hand-in-hand with the mechanism proposed by Christian which we have already discussed.

The second mechanism suggested by Wynne-Edwards is the use of various kinds of group activities which serve to indicate the size of the population and, presumably, to reduce reproduction when the population becomes too high. Many animals perform group flights, swarms, and dances, have choruses, etc., which may serve this function. The evening flight of huge numbers of starlings, the dawn and evening chorus of birds and frogs, the swarming of gnats and whirligig beetles are all cited as possible candidates for this function. At present this remains in the stage of an interesting hypothesis awaiting confirming evidence; to the extent that it requires "group" selection (see pages 95, 96), the burden of the proof lies with Wynne-Edwards.

Regulation by predators or parasites is another possibility. The explosive increase of some species when their predators were removed shows that they had been regulated by their predators. The removal of mountain lions was followed by a great increase in populations of deer on the Kaibab plateau of Arizona. Insects and plants that reach pest proportions are usually those introduced into a new continent without their predators or parasites. Prickly pear cactus was introduced from South America into Queensland and soon spread over millions of acres, excluding cattle. When a caterpillar that lived on the plant in South America was introduced into Queensland, it quickly destroyed most of the cactus (Fig. 5-17). We do not know whether these are special cases, but it is obvious that some populations are regulated by predators.

Parasites, unlike predators, cannot afford to kill their hosts, because the parasites themselves are then eliminated. If some parasites, such as disease microorganisms, kill their hosts, it is obvious that they have not achieved a balanced equilibrium. The protozoan *Trypanosoma,* which causes sleeping sickness, lives, apparently without causing harm, in the native ground-dwelling mammals in Africa. Men and cattle, which have arrived in Africa more recently, are killed by this parasite. However, we do not know what the effect might be if we removed the parasite from the native mammals. All disease, in fact, can be viewed as the consequence of ill adaptation on the part of the parasites.

At the beginning of this discussion we made a distinction between

Fig. 5-17. (a) Dense prickly pear (*Opuntia inermis*) prior to insect attack, Chinchilla, Queensland, October 1926. (b) The same area in October 1929, showing almost complete destruction by *Cactoblastis*. (Courtesy J. Mann, Director of Biological Section, Department of Lands, Queensland, Australia.)

141

organisms at the mercy of their physical environment and those that are independent of it. However, most organisms are somewhere between these extremes, and for them the degree to which the physical environment or the interactions among organisms influences the size of their population remains an open question. In fact, there is every reason to think that the influences may interact; a malnourished animal is more susceptible to predators and bad weather. Which of these shall we say is responsible for its death?

One of the reasons for our lack of knowledge is the difficulty of gathering information about populations. Measurements of physical factors, such as temperature, humidity, salinity, etc., are relatively straightforward, and therefore many studies of the effects of physical factors have been made.

In contrast, the interactions among organisms are often subtle and difficult to measure. Where they are most obvious, as in a coral reef, they are also most complex, being imbedded in a Gordian knot of relationships, which is very difficult to unravel. Two approaches to this problem have been followed. In one, the whole complex is studied intact, by using field experiments such as the exclusion of certain organisms from the interaction. The other approach (reminiscent of Alexander the Great slicing the Gordian knot with his sword) is to pluck the organisms out of their natural habitat and study the interactions in the laboratory. Both approaches are probably necessary. One of the problems with the second method is that most of the organisms that have been successfully studied in the laboratory have come from temporary, somewhat unstable environments: beetles from stored grain, water fleas from ponds, flies from carcasses and garbage cans, and mice from houses have contributed most of what we know about laboratory populations. It is possible that we may not be able to isolate species that live in stable, complex associations, and therefore other methods may have to be devised. It is clear that there is no *one* mechanism which determines the size of populations.

6

Population Interactions

Charles Darwin was one of the first to recognize the importance of competition among populations. In *On the Origin of Species* (1859, pages 77–79), he said:

> Look at a plant in the midst of its range, why does it not double or quadruple its numbers? We know that it can perfectly well withstand a little more heat or cold, dampness or dryness, for elsewhere it ranges into slightly hotter or colder, damper or drier districts. In this case we can clearly see that if we wish in imagination to give the plant the power of increasing in number, we should have to give it some advantage over its competitors, or over the animals which prey on it. On the confines of its geographical range, a change of constitution with respect to climate would clearly be an advantage to our plant; but we have reason to believe that only a few plants or animals range so far, that they are destroyed exclusively by the rigour of the climate. Not until we reach the extreme confines of life, in the Arctic regions or on the borders of an utter desert, will competition cease. The land may be extremely cold or dry, yet there will be competition between some few species or between the individuals of the same species, for the warmest or dampest spots.
> It is good thus to try in imagination to give to any one species an advantage over another. Probably in no single instance should we know what to do. This ought to convince us of our ignorance on the mutual relations of all organic beings; a conviction as necessary, as it is difficult to acquire. All that we can do, is to keep steadily in mind that each organic being is striving to increase in a geometrical ratio; that each at some period of its life, during some season of the year, during each generation or at intervals, has to struggle for life and to suffer great destruction.

COMPETITION

When the ratio of population size to essential resources reaches the point at which mortality balances reproduction, the population must stop growing. In the presence of another organism which uses any part of the same essential resource, the population of each would be smaller than if it were growing alone. This is an operational definition of the existence of competition among populations.

What are the consequences of competition between two species? Because the two species, to be distinguishable, must be somehow different, it is inconceivable that under a given set of conditions they could be equal in birth and death rates or in their effects on each other. Therefore, one of the species must be superior either in having a greater capacity to increase (r of Chapter 5) when isolated or in its ability to interfere with the r of the other species (Gause, 1934).

We discussed the per capita rate of population increase (birthrate plus immigration rate minus death rate minus emigration rate) in Chapter 5, where we called this value r. In calculus terms r was defined by $(1/N)(dN/dt)$, where N is the population size and t the time. We discussed the kind of population growth and age distribution that would result if r remained constant, and we pointed out that this would produce exponential growth rapidly exceeding all limits. Eventually, of course, such unimpeded growth always runs into limits of one kind or another and, as we pointed out, r decreases to zero, indicating that the population stops increasing. Various factors could serve to reduce r, but it suffices to see that, as population density increases, r decreases. (As a plausible mechanism, the increased population depletes the resources, thereby reducing the birthrate and increasing the death rate.) In competition, it is not the r of unimpeded growth which is of interest, but rather the r which occurs when the environment is saturated with individuals. Although a single uniform species would, by its own population increase, reduce its own r to zero, at which point there would be no further change in population, a nonuniform population (e.g., two ages, two genotypes, or two species) will not react in so simple a fashion. Here we shall treat the two-species case; for convenience we name them species A and species B. Suppose, now, that a few of both A and B are introduced into an environment. Initially both will be scarce and their growth unimpeded. However, as their combined populations increase and the r of each decreases, there comes a first time when the r of one of the species (say A) drops to zero. The interesting process begins at this stage: B still has positive *r* and increases further; this increase reduces *A*'s *r* further and thereby makes it negative; thus *B* increases while *A* decreases. This is the mechanism of competition, in which we say that "*B* has a competitive advantage over *A*." *B* is thus increasing at the expense of *A*, which is becoming rare. *If B maintains its competitive advantage even when A is rare, B will completely eliminate A. If, on the other hand, A gains the competitive advantage when rare, and B maintains the advantage when it is rare, neither can eliminate the other and both will coexist indefinitely.* Why should each species gain a competitive advantage when rare? This is essentially a difficulty of the imagination: ecologists have thought of several reasons, but there is no guarantee that the list is complete. If, for example, each competing species has slightly different resource requirements, each is slightly better off when rare relative to the critical resource supply than when common, and coexistence will result. We discuss this, and other, means of ensuring coexistence below, but here we only wish to point out that all such mechanisms so far imagined seem to rely on heterogeneity of the environment (in this case a heterogeneous resource supply).

Although this theory may seem oversimplified, many competition experiments have been performed, and, in every case, given *simple uniform* conditions and a long enough span of time, one species was eliminated. These are among the best-documented experiments of population biology. Gause (1934) with protozoa, Frank (1952, 1957) with *Daphnia* and other water fleas, and Crombie (1946) and Park (1948, 1954) with flour beetles have demonstrated this "competitive exclusion" in the bottle populations in the laboratory (Figs. 6-1, 6-2).

Of course, if the environment is not completely uniform in space one species may not be eliminated and both may coexist indefinitely. In one of his experiments, Crombie placed short lengths of fine-glass tubing in the flour. The larvae of the smaller species of beetle could crawl inside and pupate; they were protected there from attack by the other species, which were too big to get in (Fig. 6-1). Therefore, both coexisted.

If the environment changes in time, a somewhat prolonged co-existence of competitors is also possible. If, in our first experiment, conditions were changed in the course of the experiment, so that the *r* of *A* became greater than that of *B*, the direction of change of the proportions of the two species would be reversed. If the period between such fluctuations is shorter than the time required to eliminate one of the species, two species will coexist, at least for some time, although to persist indefinitely is more difficult. We shall return to this subject later.

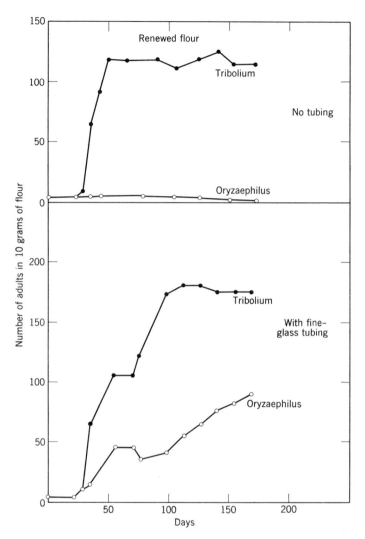

Fig. 6-1. Tribolium outcompetes *Oryzaephilus* in renewed flour, unless fine-glass tubing is added to the medium which allows coexistence. [After Crombie (1946).]

What is the evidence of competition in natural populations? Direct evidence of competitive exclusion is difficult to obtain, partly because of the difficulty of measuring numbers, mortality, etc., in most natural populations. Also, because the population being eliminated is less fit, we would expect natural selection to have already eliminated the losing competitor sometime in the past.

Let us look at one example of the direct observation of competitive exclusion of one species by another. In Scotland, one species of barnacle, *Chthamalus stellatus*, occurs in the high part of the intertidal seashore, whereas another barnacle, *Balanus balanoides*, occurs lower down. Although young *Chthamalus* often attach to the rock in the lower "Balanus" zone, after their short stay as drifters in the plankton, no adults are ever found there. The history of a population of barnacles can be recorded very accurately by holding a pane of glass over a patch of barnacles and by marking with glass-marking ink each spot where a barnacle is. Once attached, barnacles are fixed, so that by returning later we can check exactly which barnacles have died and which new ones have arrived.

By doing this, Connell (1961*b*) was able to show that in the lower zone *Balanus*, which grows faster, ousts *Chthamalus* by actually prying it off the rocks or growing over it. When *Chthamalus* was isolated from contact with *Balanus*, it lived with no difficulty in the lower zone, showing that the reason for the restriction of *Chthamalus* to the high-shore levels was interspecific competition with *Balanus* (Fig. 1-24).

There is indirect evidence that competition may have been important in natural situations. If similar species whose ranges had been

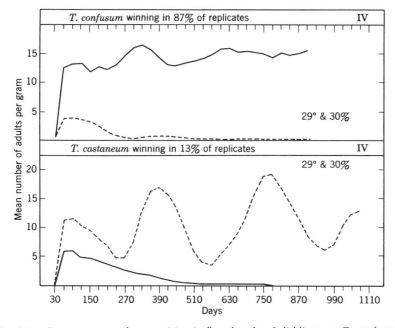

Fig. 6-2. Consequences of competition in flour beetles. Solid lines are *T. confusum;* broken lines, *T. castaneum.* (Park, 1954).

separated previously were to meet, we might expect one species to eliminate the other in competition. If, however, there were enough variation within each population and in their shared resource, we would expect that certain individuals of one species might be superior in competing for one variety of the resource, whereas certain individuals of the other species would be superior in using the opposite extreme of the resource. For example, let us imagine two very similar species of birds invading an island. Suppose that species *A* had bills ranging in size from 1.0 to 2.0 cm and species *B*, 0.5 to 1.5 cm. If they both ate any beetles they found, it seems reasonable that the large individuals of *A* would be better able to increase their population by

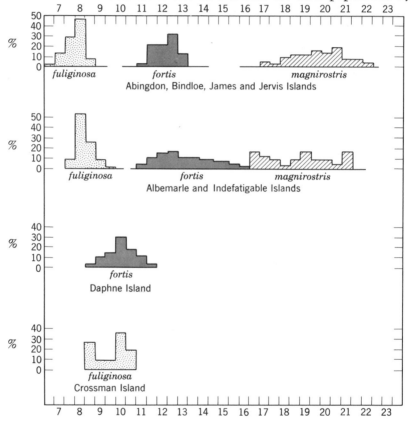

Fig. 6-3. Histograms of beak-depth in *Geospiza* species (Darwin's finches). Measurements in millimeters are placed horizontally, and the percentage of specimens of each size vertically. If the beak-depths on Daphne and Crossman islands are indicative of optima in the absence of close competitors, the displacement on islands with competitors is most easily interpreted as character displacement. The islands are in the Galapagos. (Lack, 1947.)

eating large beetles, whereas the small individuals of B would do better on small beetles. As we showed in Chapter 3 (Figs. 3-1, 3-2) there should be a limiting similarity in coexisting species. That is, similar resources are likely to support one intermediate population, rather than two discrete populations. Thus the two bird species, in order to coexist, must diverge in bill size, in the region of overlap (Fig. 6-3).

Turning the situation upside down, we could look for populations of similar organisms whose ranges partly overlap, to see whether their characteristics differed more when they occurred together than when they were apart. Because the most likely cause of such "character displacement" is competition, any instances we find are probably evidence that interspecies competition occurred in the past. In Fig. 6-4 we have shown an instance of character displacement in a bird. Where the ranges of the two species overlap, the color pattern of the

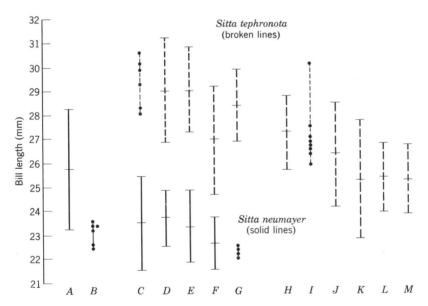

Fig. 6-4. Character displacement in Asiatic nuthatches. Bill length and facial stripe in the two species are very different in areas where they occur together, but are quite similar where they occur alone. Populations west of the zone of overlap (Sitta neumayer): A, Dalmatia and Greece; B, Asia Minor. In the zone of overlap: C, Azerbaijan and northern Iran; D, Kermanshah; E, Luristan and Bakhtiari; F, Fars; G, Kirman. East of the zone of overlap (Sitta tephronota): H, Persian Baluchistan; I, southern Afghanistan; J, Khorasan; K, north-central Afghanistan north of the Hindu Kush; L, northeastern Afghanistan (Pamirs); M, Ferghana and western Tian Shan. (Vaurie, 1951.)

plumage and length of bill is displaced. Other instances of character displacement have been found in mammals, frogs, fishes, crabs, and insects (Brown and Wilson, 1956).

Both of these examples from natural populations have shown that competing populations can coexist, if there is enough variation in the resources to allow each population to specialize in one variety of the resource. Of the two barnacles, one specialized in occupying the space at the high zone of the shore, the other on the lower zone.

If there is variation in time, a species might specialize by utilizing a temporarily abundant resource *before* another, more efficient, competitor arrived and thus gain a temporary respite from extermination. For example, we gave an illustration in Chapter 1 of how, in deciduous woodlands in Britain, the short plants became active in early spring before the trees acquired their leaves. If the climate were summerlike all year so that the trees kept their leaves, the lower plants might be shaded out of existence; but because of the seasonal variations the shorter plants could exist by narrowing their activities into a short period in early spring. In this case, the lower plants, by being immune to the competition of the trees in the summer, manage to persist indefinitely.

Another method of such "opportunistic" specialization is shown by many "weedy" plants. New ground is frequently laid bare by landslides, floods, changes in the course of rivers, the uprooting caused by falling trees, etc. Some plants specialize in colonizing these areas. They often produce many seeds, which have structures ensuring wide dispersal; they germinate and grow quickly in full sunlight and produce seeds in one season. Even if other species, which arrive later or grow slower, may exclude them in the long run, these "opportunistic" species can persist in the region, provided that the environment is continually disturbed and thus supplies new bare sites.

Thus we would expect to find potentially competing species coexisting if the environment varied in space and time. As a corollary, we would not expect such coexistence under constant, uniform conditions, unless one or both of the following conditions existed.

(*a*) If some environmental factor kept the populations of both species so low that no essential resource was limiting, no competition would occur. For example, if storms were continually occurring so that barnacles were rubbed off the rocks at such a rate that space was never limiting, competition for space obviously would not occur. The factor keeping down the populations would need to be independent of the resources. Lack of water reduces populations of desert plants but does not reduce competition; the rather even "orchard-like" pattern of desert bushes is evidence of severe competition among them.

(b) If each of the species in competition was favored when rare, and was placed at a disadvantage when common, neither would be eliminated as we pointed out above. Thus, when a species was losing in competition, getting gradually more rare, it would be progressively favored over the winning species which, in becoming commoner, was losing its advantage.

At first, it is hard to see why a species should be at an advantage in competition, the more rare it becomes. Two possible explanations seem reasonable. One explanation is clearer if we restate the situation as follows: "If each species limits its own further increase more than the others, coexistence can occur." Thus, if each species of tree required a different nutrient, such as a trace element, the growth of each species' offspring might be inhibited directly beneath it. The roots of the adult tree would be able to remove all of the essential nutrient from the upper soil; other species, which did not require this nutrient, would be able to grow in this space.

The second explanation depends on the presence of another organism in the system—an "enemy" such as a predator or parasite. A characteristic of a predator or parasite is that it must be flexible in its choice of prey. If one species of its prey becomes scarce, it must be able to change to another, more abundant, kind of food. Let us imagine two species of prey organisms competing with each other and both being attacked by a predator. As one competitor begins to lose, it becomes progressively scarcer. Eventually the predator will cease feeding upon it and will turn to another prey; whether this other species is the other competitor or a completely different species is irrelevant. The point is that the rare species has gained an advantage by being rare, and thus will increase its population instead of being eliminated.

One of the first examples of the coexistence of competing species because of attacks by predators was seen by Charles Darwin (1859, op. cit., pages 67–68):

If turf which has long been mown, and the case would be the same with turf closely browsed by quadrupeds, be let to grow, the more vigorous plants gradually kill the less vigorous, though fully grown plants; thus out of twenty species growing on a little plot of mown turf (three feet by four) nine species perished, from the other species being allowed to grow up freely.

Since then there have been other instances where grazing animals have been excluded from vegetation; the result has generally been the elimination of some species by competition as the plants grew taller (Summerhayes, 1941). When predatory snails were kept from eating the barnacles in the study already described (Connell, 1961b), competition was more severe; however, the predation evidently was not sufficiently effective to prevent exclusion of *Chthamalus* by *Balanus*. In another study, where two species of predatory snails were feeding on barnacles, they reduced the populations so much that no competition occurred at all; when the snails were excluded for several years, one species of barnacle eventually occupied the area, excluding other barnacles and also the grazing limpets (Connell, 1964). A similar thing happened when Paine (1966) removed predatory starfish from a stretch of intertidal shore; barnacles first occupied most of the area and were later crowded out by mussels; grazing chitons and limpets were also crowded out. Thus, evidence has been accumulating that competition is reduced and coexistence of many species favored if the competitors are attacked by predators and parasites. Let us look at these interactions in more detail.

PREDATOR-PREY INTERACTIONS

Each individual predator will strive to get as much food as it can. Therefore, we would expect natural selection to increase the efficiency of the predator in finding and eating prey. In similar manner, selection would favor any individual prey which is able to escape being eaten. (Even if it is not killed, the amount consumed by a parasite or grazing herbivore reduces the prey's ability to survive and reproduce.)

These two selective forces act in opposite directions; as the prey evolves traits that increase its ability to avoid being eaten, the predator evolves more efficient means of eating them. In evolution, as in real life, the prey is just one jump ahead of the predator!

We shall introduce a graphical method of dealing with predator-prey interactions (Rosenzweig and MacArthur, 1963). If we draw a graph with prey density as abscissa and predator density as ordinate, each point will represent a unique combination of the two populations. Let us now, in imagination, set up many experiments starting at many different densities of each, and record whether the prey and predator populations increase or decrease at each point. We can insert two arrows at each point, showing the direction of change of prey and predator. The resultant of the two is a single arrow at each point. Figure 6-5 illustrates a possible graph at this stage.

Now suppose (and this is a very important supposition) that the direction of population change, at successive times from the same

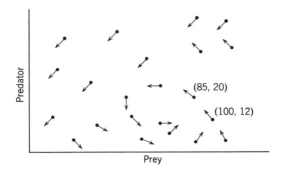

Fig. 6-5. Hypothetical population changes in sample populations of predator and prey. For further discussion, see text.

point, is always the same. This is equivalent to supposing that either the environment is not changing except as a consequence of the population changes under consideration, or that it is subject only to changes that do not affect the population changes of prey and predator. For example, we are assuming that 100 prey and 12 predators, under the conditions we are describing, will always change toward 80 prey and 15 predators. We have to make this assumption, so that the arrows on the graphs have a general significance and do not simply refer to what happened once. If we cannot say that the change always proceeds in the same direction (because the varying population of a third species affects how the first two will change for example), we have to abandon, or at least modify, the application of the graphs. We shall use this admittedly oversimplified approach and shall later compare our findings with some real-life situations. If we find a close correspondence, we can say that the assumption of independence of outside effects is justified in some real-life situations and therefore use the theory.

At certain points we find that the prey population does not change. By joining these points we could draw a curve separating a zone in which prey increases from one in which it decreases. We can deduce the general shape of this curve as follows. When there are no predators, there is a population size above which the prey population must decrease because of shortage of food (or of space or any other requisite in short supply). Similarly, there is a lower boundary below which the population would decrease, because contacts between the sexes would be too few to maintain the population. Thus the curve must intersect the abscissa at two points.

In the presence of predators, the curve must slope up to the right from the lower limit; because prey are being eaten, more prey will now be necessary for successful reproduction. From the upper limit, the curve must slope back to the left: at the upper limit of resources, some of the prey population is being used to feed the predators; the predators are simply an additional onus on the resources. Thus the two ends of the curve slope upward toward each other, and the entire curve forms a single peak, as shown in Fig. 6-6; below the curve the prey increases, above the curve it decreases.

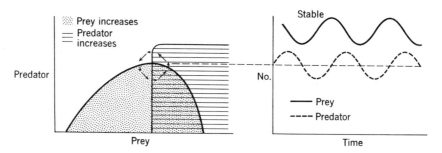

Fig. 6-6. Predator and prey zones of increase, corresponding to Fig. 6-5, are shown at the left. At the right is the corresponding time graph.

The shape of the predator curve can be deduced as follows. When there are relatively few predators, the hunting effort and effectiveness of the predator is almost independent of the number of other predators. Thus the rate of food gathering depends mostly on prey density. When the prey fall below a certain density, the predators decrease; when the density is higher than this level, the predators increase. Thus the curve of no predator increase is approximately vertical. (This does not mean that a large number of predators will not deplete the prey population faster than a small number. It does mean that at any instant the rate of catching per predator depends only on the density of its food supply.) If, however, the predators need to accumulate much food before reproducing, a large predator population will need a denser prey population in order to maintain itself. In this case, the predator line will slope up and to the right.

We can now draw arrows on the graph to see how the populations will change (see Fig. 6-6). Although we shall be more precise later, we can already see that the chain of arrows leads around back toward its beginning, each species being alternately rare, then common, then rare, and so on. These changes are usually called predator-prey oscillations. Before commenting on examples in nature and in the laboratory, however, we must analyze this graph more deeply. For example, if the arrows lead precisely back to their beginning, the course of population changes will repeat itself exactly over and over again (Fig. 6-6). If, on the other hand, the arrows do not form a closed path, but spiral *inward* (Fig. 6-7), the magnitude of the oscillations will decrease and "damped" oscillations and eventually a steady-state population of predator and prey will develop. If the arrows spiral *outward* (Fig. 6-8), the magnitude of the oscillations will increase and, quite likely, the oscillations will become so large that the arrows will intersect the ordinate in the graph (i.e., the predator will exterminate the prey), and then, of course, the predators will also vanish.

If the relative magnitude of change in the two populations were the same at all points not on the lines, we could draw the arrows at 45°. This is not strictly the case, of course, but for heuristic purposes we treat 45° as an "average" slope. By doing so, we can predict whether the interaction will be stable or unstable. If the vertical predator line is to the left of the prey peak, so that the intersection point is on the part of the prey curve that slopes down to the left, the spirals are outward and are ever-increasing. If the predator line is to the right

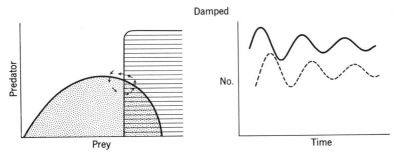

Fig. 6-7. Legend as in Fig. 6-6. Here the predator's zone of increases intersects the descending part of the prey's, causing damping of the oscillations.

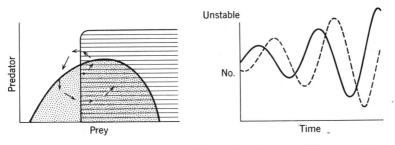

Fig. 6-8. Legend, as in Fig. 6-6. Here the predator's zone of increase intersects the ascending part of the prey's, causing increasing oscillations.

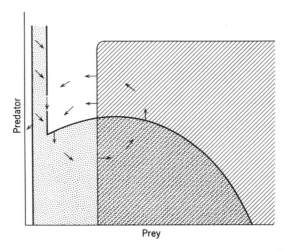

Fig. 6-9. As in Fig. 6-8, but prey hiding place puts a limit to the increase of the amplitude of the oscillations. The predator ceiling, too, could limit the oscillations.

of the prey peak, where the prey line slopes down toward the right, the spirals are inward and are decreasing. (You can verify this by drawing arrows at 45° on a piece of scratch paper.) That is, if the predator can only increase when the prey are so common that their food is becoming short, oscillation will be of the stable, inward type (Fig. 6-7). If, on the other hand, the predator is so efficient at catching the prey that it can increase when prey are very scarce, oscillations are likely to be unstable, outward spirals (Fig. 6-8). This is a striking case of a common phenomenon—processes carried out with the greatest immediate effectiveness are usually the least stable. Most laboratory examples are not of the inward spiraling type—the predator is so effective in finding the prey that it exterminates all or almost all of the prey and then starves to death (see Fig. 6-10).

If, on the other hand, *complications* are introduced into the system, such as hiding places for the prey and/or alternate foods for the predator, it seems that the arrows tend more to spiral *in*, or at least they no longer spiral out, thus producing a stable system. The complicated cases, of course, correspond more closely to what we observe in nature. We can see why hiding places for the prey should stabilize the system: a hiding place means that near the predator axis—that is, where prey are scarce—the prey can hide, and a larger number of predators is necessary to depress the prey increase. This means that the left side of the prey's boundary line is raised. If there is an inviolable hiding place, the graph may look like Fig. 6-9, with the system stabilized as shown.

In fact, it seems to be a general principle that complicated natural systems are stabler than simple ones. For example, the few mammal species present in arctic regions fluctuate radically in numbers, as the fur-catch from the Hudson Bay Company indicates (see Fig. 5-7). In milder climates, in which more species are present in a more complicated situation, such wild population fluctuations are almost unknown.

Huffaker (1958) produced a laboratory example of the effect of complicating an environment and thereby reducing predator efficiency. He used two mite species, one of which fed on oranges and the other of which was a predator, feeding on the first. When the situation allowed the predator to move easily from orange to orange, the predator found and exterminated most of the prey and then itself died out, as in Fig. 6-10.

The prey (but not the predator) could move by floating from high

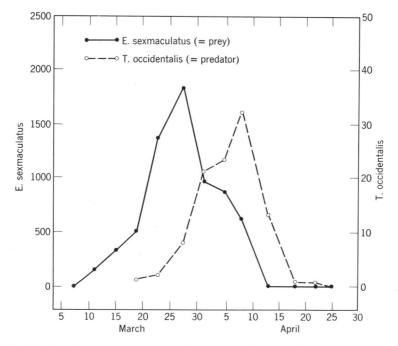

Fig. 6-10. Densities per orange-area of the prey, *Eotetranychus sexmaculatus,* and the predator, *Typhlodromus occidentalis,* with 20 small areas of food for the prey (orange surface) alternating with 20 foodless positions—a 2-orange feeding area on a 20-orange dispersion. (After Huffaker, 1958.)

points on strands of silk. When Huffaker slowed the predator movements with partial barriers and speeded up the prey movements by putting in "launching posts," the prey persisted in the face of predation. Evidently, the "fugitive" prey could build up local populations from which many prey could disperse. Predators would eventually find this local prey concentration and would exterminate it while a new prey population was increasing somewhere else. As a result, the two species persisted with periodic oscillations as in Fig. 6-11.

One additional feature of the predator-prey graphs is of interest: a ceiling on the predator population. Suppose that the predators are territorial, so that their population can never increase beyond some value. Then the graphs are modified either as in Fig. 6-12 or in Fig. 6-13, which correspond to a low "ceiling" and a high one, respectively. When the ceiling is low (i.e., when the predator population is kept below the level that the food would support), there are no oscillations at all. When the ceiling is high, it limits the size of the oscillations and may prevent the populations from going extinct. Again, the complications (of predator territoriality) help to stabilize the system.

Because complexity appears to be the basis for the stability of predator-prey systems, it is not surprising that their interactions are not yet understood. Huffaker made his systems very complex, but nevertheless obtained rather large oscillation when the predator and prey persisted together. When several species of both predator and prey interact, stability may be further increased. One species of prey may support more than one species of predator; certainly, one species of large tree supports many species of insects, and one horse has many species of parasites. There may be more predatory species than herbivorous species in the insects; most of the Hymenoptera, a large fraction of the Coleoptera, most Diptera, all Neuroptera, Trichoptera, Mecoptera, and Odonata are predators; we may also add all of the arachnoids, except some mites.

Finally, we may ask whether predators have evolved efficient strategies for "harvesting" their prey. Predators should maximize the amount of prey food (which predators cannot eat) by turning it into prey tissue (which they can eat). If they reduce the prey population too much, some prey food will be left uneaten; this might then be channeled into another food chain and thus be wasted to this predator. One case has been studied in which a predator seems to have adopted an efficient management strategy. *Thais lapillus* is a snail that eats barnacles. It chooses the larger ones, probably because it gets more food per unit effort by doing so. However, this choice tends to remove selectively the prey individuals that grow at a slower rate (the older

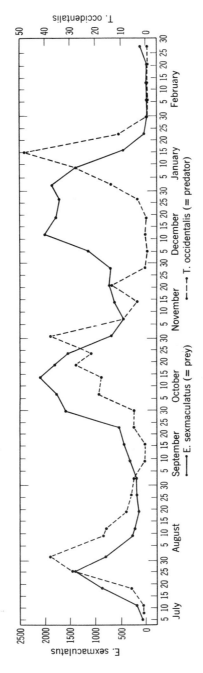

Fig. 6-11. Three oscillations in density of a predator-prey relation in which the predatory mite, *Typhlodromus occidentalis*, preyed upon the orange feeding six-spotted mite, *Eotetranychus sexmaculatus* in a much more complicated environment. (After Huffaker, 1958.)

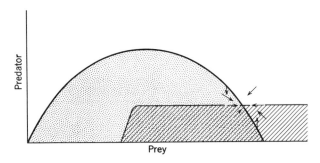

Fig. 6-12. Legend as in Fig. 6-6. Here the predator ceiling prevents oscillations altogether.

ones). In other words, *Thais* maintains a population of young fast-growing prey, which turn prey food into prey tissue faster than would a population of older prey (Connell, 1961*a*).

This graphical system for analyzing predator-prey interactions can be extended to several predators competing for two prey (MacArthur and Levins, 1964) and also to triple systems of carnivore, herbivore, and plant. We show a three dimensional carnivore-herbivore-plant graph in Fig. 6-14 for the reader who wishes to test his understanding of the simple ones.

Finally, we encourage the reader to draw and interpret graphs describing competition interactions analogous to the predator-prey graphs used here. These graphs apply both to interspecific competition and to natural selection (interallelic competition, but not for space on a chromosome; rather, for the same resources for which the individuals are competing) (MacArthur, 1962). For interest we include four of these competition graphs (Fig.·6-15) (others, combining portions of

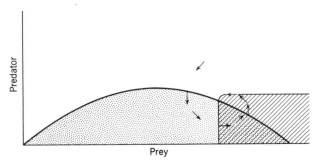

Fig. 6-13. Legend as in Fig. 6-6. Here the predator ceiling damps all oscillations to one cycle.

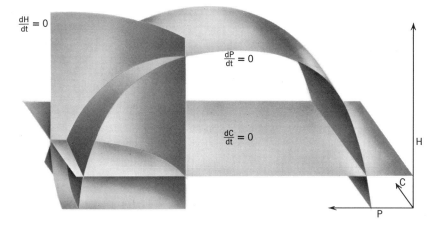

Fig. 6-14. An attempt at a three-dimensional version of Fig. 6-6. C is a carnivore, H a herbivore, and C a plant.

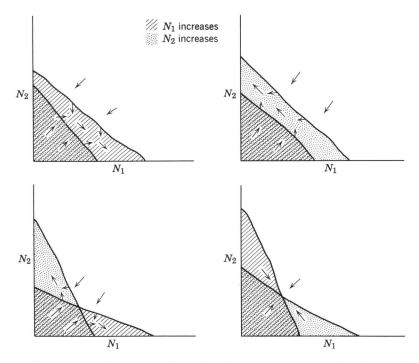

Fig. 6-15. Competition graphs illustrating the four basic competitive interactions. N_1 and N_2 can be populations either of competing species or of competing alleles. In the latter case, these graphs provide a density-dependent version of the population genetics of Chapter 2.

these, can be constructed with wiggly lines), but we leave their interpretation to the reader.

Volterra's principle. We end our study of predator-prey interactions by returning to a simpler and less realistic model which lends itself better to equations (although the results that we derive apply more widely). We suppose that the predator population, x, and the prey population, y, interact (by predator eating prey) when they meet and that this occurs proportionally to the product, xy. (This is equivalent to saying that predators search independently for prey). Furthermore, we assume that the predator rate of increase has a positive component proportional to the number of meetings that result in predator eating prey. Finally, we assume that it will decrease in the absence of food (prey). Symbolically, we can write

$$\frac{dx}{dt} = I_x xy - D_x x \tag{1}$$

where I_x and D_x are the per capita rates of increase and decrease, whose balance determines the population change. Similarly

$$\frac{dy}{dt} = I_y y - D_y xy \tag{2}$$

governs the change in prey population, for its decrease is proportional to the predation interactions that occurred with frequency xy. We could plot graphs that correspond to these equations (vertical and horizontal lines), but by doing the same mathematically we derive a remarkable result due to Volterra (1926). Here we shall not be concerned with the oscillations, but only with the equilibrium point for which $dx/dt = 0$ and $dy/dt = 0$. For the values of x and y, at which these derivatives are zero, both populations will remain constant; hence it is an equilibrium point. From Equations 1 and 2 we see, by setting the derivatives equal to zero, that there is an equilibrium when $y = D_x/I_x$ and $x = I_y/D_y$. This is not remarkable, but Volterra noticed the effect on this equilibrium of additional mortality acting both on predator and prey (e.g., extra mortality imposed by man). Consider first Equation 1, which governs the predator population. The additional mortality will not, unless very severe, change the interaction term $I_x xy$; therefore, it must enlarge the number D_x. Similarly, it will not alter the interaction term $D_y x_y$ in the equation for dy/dt; hence it must be reflected in a reduced value of I_y. How does this alter the equilibrium value $y = D_x/I_x$? It leaves the denominator unchanged and increases the numerator, so that the prey population will *increase* as the result of man's killing both predator and prey! The predator equilibrium population size, $x = I_y/D_y$, will decrease, because I_y was reduced and D_y was unaltered. Hence, Volterra concluded, *a cause moderately destructive to both predator and prey will increase the average prey population and decrease that of the predator.*

This unexpected prediction has spectacular application to insecticide treatments, which destroy both insect predators and their insect prey. Volterra's principle implies that the application of the insecticides will, unless *extremely* destructive to the prey, increase the populations of those insects that are kept in control by other predatory insects. A remarkable confirmation came from the cottony cushion scale insect (*Icerya purchasi*), which, when accidentally introduced from Australia in 1868, threatened to destroy the American citrus industry. Thereupon its natural Australian predator, a ladybird beetle, *Novius cardinalis*, was introduced and took hold immediately, reducing the scale to a low level. When DDT was discovered to kill scale insects, it was applied by the orchardists in the hope of further reducing the scale insects. However, in agreement with Volterra's principle, the effect was an increase of the scale insect! This shows the danger of tampering with those aspects of nature that are not understood (Elton, 1958).

7

Ecological Communities

INTRODUCTION

We have discussed so far either single populations or the interactions of a few populations. In natural conditions, however, many hundreds of species may live in close proximity. Each population is dependent on the activities of many others. Let us illustrate this by considering what might happen to a species isolated from all others. If it were an "animal" species, that is, one which could not trap energy from a physical source such as light or from inorganic molecules, it would starve. (Notice that by this definition fungi and many bacteria are called "animals.") However, what about the photosynthetic bacteria and green plants, and the chemo-synthetic bacteria? It certainly must be possible to rear them in completely "inorganic" conditions isolated from other species, because all they need are an inorganic energy source and a supply of inorganic matter for use in building the necessary structures. For photosynthesizers the supply of solar energy can be considered essentially unlimited, but not the supply of inorganic matter. As an example, suppose that we isolated a species of green plant in an otherwise sterile field and allowed its population to grow. If this plant species were a very tall tree, it might consume the entire supply of essential nutrients that its roots could reach in the soil. All chemical elements in the soil that are used by plants would then be incorporated into the tissues of the trees. (This example is not so farfetched as it might seem. In tropical rain forests, the soil often has few useful mineral nutrients, because rain washes very small molecules away. If the trees are re-moved, one or two seasons of crops will exhaust the few remaining nutrients.)

At this point growth would stop for lack of materials. When some of the leaves or other tissues died and were broken up into molecules small enough for the roots to absorb, further growth would be possible.

This would be a very slow process in our sterile field; still undecom-posed plant remains have been found in peat and coal deposits of great age, despite evidence of microbial action. In our example, there were no organisms such as boring insects, bacteria, or fungi which normally break up dead plants. Only weathering and physical breakdown of the large molecules would occur. Under these unusual conditions, it is dif-ficult to predict what might happen, but the trees would almost cer-tainly die and fall to the ground because of inevitable accidents, storms, etc. Since decomposition in the sterile field would be relatively much slower than it is in the real world, the field would eventually become a jumble of dead trees, with a few living spindly individuals growing at a rate determined by the rate of supply of inorganic matter released by the physical decomposition of the dead trees.

In contrast, we have actual forests, where species of animals and of decomposers, such as fungi and bacteria, attack the dead plants, breaking them into molecules small enough to be absorbed into the roots of the trees. The trees in such a forest would obviously leave more offspring than a species which resisted the attacks of all decomposers; the latter would suffer the fate of our hypothetical isolated tree species in a sterile field.

In general, then, organisms living in communities of many interacting species are fitter than if they existed in isolated populations. Moreover, because these communities are composed of organisms that are inter-dependent, they possess characteristics which emerge as a result of the interactions between the different species. Thus we may study them as units, just as we studied cells, organisms, and populations of a single species.

GROWTH AND DEVELOPMENT OF A COMMUNITY

Patches of bare ground both on land and under water are continually formed by landslides, lava flows, changes in the course of rivers and in the level of the sea, etc. They never remain bare for long; with the ex-ception of the highest mountains, polar regions, and extreme deserts, the surface of the land is covered with vegetation. Most of the sea floor also is well occupied, although more sparsely at greater depths and in such unstable places as sandy beaches.

The question we would like to investigate is whether the process of colonization of a bare patch is haphazard or predictable. Are the species which eventually compose the community those that happen to arrive and occupy the ground first, or is there a predictable pattern of development leading to a particular set of organisms in any one place? Let us begin with an actual example.

When a farmer in the eastern part of the United States abandons a cultivated field, it may appear bare for a few weeks or months, but it is soon covered with a carpet of small plants. As the years progress, some remarkable changes take place. Small annual plants are replaced by larger perennials, and soon clumps of bushes and young trees appear. In New England, the first trees may be pine, spruce, or birch; many years later the forest that once was a field will have changed its composition and sugar maple, beech, and hemlock may become important. At this stage, changes take place very slowly, and it is useful to consider the limit of this process, which is completely self-perpetuating. This limit is called the local "climax" forest. The composition of the climax seems to vary continuously with soil type, climate, and other features, but it is often independent of the starting conditions of the succession. If, for example, the farmer abandoned a checkerboard of fields and severely cut woods, all on homogenous topography and soil, the initial phases of the succession would be very different. Thus blackberry bushes and young trees would play a more important role in the early stages of the cut-forest succession. Nevertheless, remarkably enough, the two successions would converge to a similar climax and the checkerboard pattern of the abandoned farm may gradually be obliterated. An example of succession in plants and animals is shown in Fig. 7-1.

Succession may also be observed in quite different circumstances. The kinds of insects that infest granaries change radically from month to month in a regular succession. Here no climax is reached, however, because the stored grain does not renew itself and does not become available again for further generations of beetles, as the soil nutrients do in the forest. Again, if we keep track of the protozoa that inhabit a jar of water in which hay has been boiled, we will discover spectacular changes in composition from day to day. Finally, succession can be distinguished in organisms that float in the ocean: when a mass of water rich in nutrients wells up to the surface, it is colonized by certain floating plants and associated animals. As time progresses, the "plankton" organisms change in species composition in a regular pattern.

Clearly, there are several questions we may ask. What causes the

Time in years_ _ _ _ _ _ _ _ _ _ 1–10	10–25	25–100	100+
Community type_ _ _ _ _ _ _ _grassland	shrubs	pine forest	hardwood forest

	1–10	10–25	25–100	100+
Grasshopper sparrow	———			
Meadowlark	———			
Field sparrow		———— - -		
Yellowthroat		———		
Yellow–breasted chat		———		
Cardinal		———————————————		
Towhee		———————————— -		
Bachman's sparrow		———		
Prairie warbler		——		
White–eyed vireo		- ———— - - -		
Pine warbler		———		
Summer tanager		- —————————		
Carolina wren		—————————		
Carolina chickadee		——————		
Blue–gray gnatcatcher		———————		
Brown–headed nuthatch		- ——		
Wood pewee		—————————		
Hummingbird		—————————		
Tufted titmouse		—————————		
Yellow–throated vireo		—————————		
Hooded warbler		—————————		
Red–eyed vireo		- - ———————		
Hairy woodpecker		- - ———————		
Downy woodpecker		- - ———————		
Crested flycatcher		- ———————		
Wood thrush		———————		
Yellow–billed cuckoo		———————		
Black and white warbler		———————		
Kentucky warbler		———————		
Acadian flycatcher		—————		

Number of common species[a]	2	8	15	19
Density (pairs per 100 acres)	27	123	113	233

[a] A common species is arbitrarily designated as one with a density of five pairs per 100 acres or greater in one or more of the four community type.

Fig. 7-1. The general pattern of secondary succession on abandoned farmland in the southeastern United States. The upper diagram shows four stages in the life form of the vegetation (grassland, shrubs, pines, hardwoods), whereas the bar graph shows changes in passerine bird population that accompany the changes in autotrophs. A similar pattern will be found in any area where a forest is climax, but the species of plants and animals that take part in the development series will vary according to the climate or topography of the area. (After Odum, 1959.)

changes in composition? What determines whether there will be a climax? Why do different initial situations lead to the same climax? What other phenomena accompany the changes in species composition? We shall try to answer these questions.

In part, the forest succession is artificial. That is, trees become conspicuous later than annual plants, because it takes more years for a tree seed to become a conspicuous tree; if we examine a field in the early stages of succession, we often find many young tree seedlings. In this case there is little real succession; the trees were always there. This is not the whole story, even for the forest succession, and it is not at all true for other successions in which the life-span of an organism is short compared with the time required to pass through the stages of succession. Why, then, does one group of grain-infesting insect species replace another? Certain species of insects break up the whole grains into flour. The flour tends to clog their spiracles and so they die out, to be replaced by others that are adapted to living in fine flour. This gives us a clue to all of the true replacements of succession: *each species alters the environment in such a way that it can no longer grow so successfully as others.* The others replace the first species, and themselves alter the environment farther, until they are replaced by others. In the case of forest succession, each species is able to stand deeper shade than the previous one, and as the forest grows the canopy becomes thicker and casts an even deeper shade. In this new, deeper shade other species are more successful. Foresters have tables (see Table 7-1) of "tolerance" of different tree species; tolerant species are those that are successful in shade. As expected, the climax forests are composed of the most tolerant species. It should not be inferred that tolerance is merely the ability to carry on photosynthesis in the shade; it seems to be more connected with the plant's ability to compete for soil moisture in the shade. It is better just to call it a measure of the plant's success in the shade.

Next, let us inquire why a climax is reached. Probably the answer is simple: there are no more successful species to replace the climax ones. Species that are climax in one region may be replaced by even more tolerant species in other regions, showing that there is nothing unique about climax species. Why do different initial conditions lead to the same climax? This is a mathematical property of replacement processes, but the purely biological explanation is that replacement (in tree species) imparts a premium on dense canopy and tall trees (to compete successfully in the shade.) As long as both initial conditions are such that the succession replacement is directed toward more and more shade, it follows that both are likely to end up with the same shade-tolerant tree species.

Table 7-1 *"Tolerance" to Shading of American Forest Trees, from Baker (1950).*

(Cases of great uncertainty marked with*)

Eastern Conifers

Very tolerant:
  Eastern hemlock
  Balsam fir
  Atlantic white cedar
Tolerant:
  Red spruce
  Black spruce
  White spruce
  Northern white cedar
Intermediate:
  Eastern white pine
  Slash pine

Bald cypress
Intolerant:
  Eastern red cedar
  Red pine
  Pitch pine
  Shortleaf pine
  Loblolly pine
  Virginia pine
Very intolerant:
  Tamarack
  Jack pine
  Longleaf pine

Eastern Hardwoods

Very tolerant:
  Eastern hop-hornbeam
  American hornbeam
  American beech
  American holly
  Sugar maple
  Flowering dogwood
Tolerant:
  Red maple
  Silver maple
  Box elder*
  Basswood
  Tupelos
  Persimmon*
  Buckeyes
Intermediate:
  Yellow birch
  Sweet birch
  American chestnut
  White oak
  Red oak
  Black oak
  American elm
  Rock elm
  Hackberry

Magnolias*
White ash
Green ash
Black ash
Intolerant:
  Black walnut
  Butternut
  Pecan
  Hickories
  Paper birch
  Yellow poplar
  Sassafras*
  Sweet gum
  Sycamore*
  Black cherry
  Honey locust
  Kentucky coffee tree
  Catalpas
Very intolerant:
  Willows (as a class)
  Quaking aspen
  Bigtooth aspen
  Cottonwoods
  Grey birch
  Black locust
  Osage orange

Western Conifers

Very tolerant:
  Western hemlock
  Alpine fir
  Western red cedar
  Pacific yew
  California torreya
Tolerant:
  Sitka spruce
  Engelmann spruce
  Mountain hemlock
  Pacific silver fir
  Grand fir
  White fir
  Redwood
  Incense cedar
  Port Orford white cedar
  Alaska yellow cedar*
Intermediate:
  Western white pine
  Sugar pine
  Monterey pine
  Blue spruce

  Douglas-fir
  Red fir*
  Giant sequoia
Intolerant:
  Limber pine
  Piñon pines*
  Ponderosa pine
  Jeffrey pine
  Lodgepole pine
  Coulter pine
  Knobcone pine
  Bishop pine
  Big-cone spruce
  Noble fir
  Junipers
Very intolerant:
  Whitebark pine
  Foxtail pine
  Bristlecone pine
  Digger pine
  Western larch
  Alpine larch

Western Hardwoods

Very tolerant:
  Vine maple
Tolerant:
  Tan oak
  Canyon live oak
  Big-leaf maple
  Madrone*
  California laurel *

Intermediate:
  Red alder*
  Golden chinquapin*
  Oregon ash
  California white oak*
  Oregon white oak*
Very intolerant:
  Quaking aspen
  Cottonwoods

There are exceptions and variations in this scheme, which illuminate its operation. In harsh environments such as extreme deserts, where the permanent vegetation consists of widely spaced bushes, the first colonizers are often these same species. In contrast, where conditions are more favorable, as in forests, the first species to colonize a bare area are hardly ever those which constitute the climax community.

The explanation of these differences is a simple one. In the desert, the presence of low, widely spaced individuals does little to modify the physical environment. To a colonizing seed of a bush there is little difference between an area with or without adult bushes. Most of the space above ground is unoccupied in either case.

In contrast, the trees of a dense forest modify the environment, making it darker, damper, less windy, etc. Unlike the desert, an entirely different set of germination requirements is produced by removing the climax vegetation. Thus the degree of succession expected is proportional to the degree of modification of the habitat by the climax species.

An interesting example of these differences occurs in Australia where dense, dark rain forest joins the bright, sparse, *Eucalyptus* forest. *Eucalyptus* seedlings obviously must be able to germinate and grow in the bright, relatively dry conditions of these open woodlands; the reverse is true of seedlings of the rain forest species. Along the border where they meet, the "light-demanding" trees of the early stages of development of a rain forest germinate and grow, forming shade under which the climax species of the rain forest can become established. Thus the rain forest advances into the *Eucalyptus* forest; within the border of the rain forest a few large, old *Eucalyptus* persist as evidence of the advance. The rain forest makes its own environment as it advances, so the juncture of the two forests is not gradual but very abrupt; entering the rain forest is like plunging into a tunnel.

The development of a community from bare ground to the equilibrium state in which the species are replacing themselves is analogous to a cell or organism reaching the stage in which it can reproduce. In this analogy the different tissues of the organism represent the different species in a community. In the embryo, the presence of a tissue determines the fate of other tissue to follow; in a developing community, the presence of a species likewise determines the species that follows.

At this point we may ask what part of the earth is covered with mature "climax" communities? This depends on the rate of disturbance. Man, the great destroyer, keeps much of the earth in a state of "early succession"; many of his crops, notably grain, are plants able to germinate and grow in plowed soil—they are early succession species. If we consider only areas undisturbed by man, there are natural disturbances, such as we described earlier, which start new successions. Both lightning fires and storms play a large role. Even within a "climax" tropical rain forest, large, old trees, weakened by the attack of many insects and fungi, are blown down in storms, uprooting the ground and smashing smaller trees. In the new opening, early succession species may sprout and grow quickly in the sun, and in their shade the climax species germinate and grow.

Because mortality is inevitable, communities, like single populations, have an age structure, with "younger" and "older" species (early and late in succession) living together. A climax community is in a dynamic equilibrium, in which the older members are continually replaced, as they die, by younger ones. With a higher rate of mortality (disturbance), a higher proportion of the community is in early stages of development (succession).

SIZE AND DIVERSITY OF COMMUNITIES

How many individuals of each species will an area support? For the time being we shall assume that populations do not change radically with time, which is true for most vertebrates and many invertebrates, and probably for most plants. We shall answer the basic problems in several steps.

First, how much life will an area support? What is the quantity of life supported by a temperate forest in rich soil with plenty of water? And how is the measurement made? In the case of large plants, there may often be three to six acres of leaf area in each acre of forest. The reader can verify or correct this estimate for any deciduous forest that he visits in the late autumn when the leaves are down. If a knitting needle pushed vertically down through the fallen leaves passes through (and therefore catches) on the average, say, 3.2 leaves, there must have been on the average 3.2 layers of leaf above any point in the forest and therefore 3.2 acres of leaf per acre of forest. No separate leaf counts and leaf-area measurements need to be made! The amount of new wood laid down each year can also be easily estimated (by taking the thickness of the year's "growth ring" and by calculating the growth geometrically; we make assumptions about proportions of wood in the main trunk.) Both the leaves and the new wood represent net production of material in the forest by all of the plant species combined. This is, of course, high-energy material, which the plants produced from the nutrients of the air and soil with energy from sunlight. Thus, if we burn the wood and leaves, reducing them again to low-energy nutrients comparable to the original, and if we do this in a calorimeter so that we can measure the amount of heat evolved, then* we can estimate

---

* This is a consequence of the first law of thermodynamics—law of conservation of energy—which, of course, implies that, no matter what the chemical pathways are, the exact amount of energy *stored* in a synthesis must be *released* in any degradation to the initial components. Otherwise, it would be possible to decompose a compound and reconstruct it by a different process and achieve a net gain in energy, thus contradicting the first law.

how much energy from the sun was stored by the plants in this high-energy form. Our answer will come out in units such as calories of energy stored per acre per year. Table 7-2 gives a summary, from various areas and parts of the earth, of the net production of plants as measured by various workers. See also Fig. 7-2.

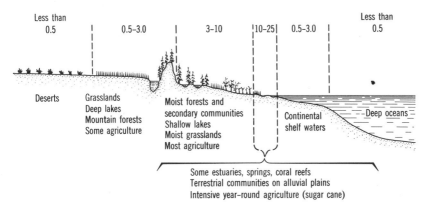

Fig. 7-2. The world distribution of primary production, in grams of dry matter per square meter per day, as indicated by average daily rates of gross production in major ecosystems. (Odum, 1963.)

A glance at the table indicates that desert regions fall far below lush fields and forest in their production. However, once they are irrigated they, too, support a very large production. Evidently, water shortage limits the production at a level lower than the nutrients and available energy could support. In the desert, the many plant species with their different sources of the scarce water (some get it deep in the ground, others store it from the occasional flash floods, etc.) can probably store more energy than could any of the species separately (although no one has demonstrated this).

As a final plant example, consider the passively drifting algae in lakes. Their numbers fluctuate greatly from month to month, and there are definite seasons of algal "bloom." As we described in Chapter 1, in temperate regions in spring and again in fall there is an overturn of the lake water with consequent thorough mixing, so that the nutrients released from the mud at the bottom are made available at the top where algae can utilize them. Some algal "blooms" follow closely after the overturn and some can be brought on artificially by fertilization with nitrogen or phosphorus. In either case, the bloom follows release of nutrients. Hence we can conclude that shortage of nutrients, rather

Table 7-2   Annual Net Primary Productivity of Various Cultivated and Natural Ecosystems as Determined by Use of Harvest Methods*

| | Net Primary Production (Grams per Square Meter) | |
Ecosystem	Per Year	Per Day
Cultivated Crops		
Wheat, world average	344	0.94 (2.3)*
Wheat, average in area of highest yields (Netherlands)	1250	3.43 (8.3)
Oats, world average	359	0.98 (2.4)
Oats, average in area of highest yields (Denmark)	926	2.54 (6.2)
Corn, world average	412	1.13 (2.3)
Corn, average in area of highest yields (Canada)	790	2.16 (4.4)
Rice, world average	497	1.36 (2.7)
Rice, average in area of highest yields (Italy and Japan)	1440	3.95 (8.0)
Hay, U.S. average	420	1.15 (2.3)
Hay, average in area of highest yields (California)	940	2.58 (5.2)
Potatoes, world average	385	1.10 (2.6)
Potatoes, average in area of highest yields (Netherlands)	845	2.31 (5.6)
Sugar beets, world average	765	2.10 (4.3)
Sugar beets, average in area of highest yields (Netherlands)	1470	4.03 (8.2)
Sugar cane, world average	1725	4.73 (4.7)
Sugar cane, average Hawaii	3430	9.40 (9.4)
Sugar cane, maximum Hawaii under intensive culture	6700	18.35 (18.4)
Mass algae culture, best yields under intensive culture outdoors, Tokyo	4530	12.4 (12.4)
Noncultivated Ecosystems		
Giant regweed, fertile bottomland, Oklahoma	1440	3.95 (9.6)
Tall Spartina salt marsh, Georgia	3300	9.0 (9.0)
Forest, pine plantation, average during years of most rapid growth (20-35 years old), England	3180	6.0 (6.0)
Forest, deciduous plantation, England, comparable to the above pine plantation	1560	3.0 (6.0)
Tall grass prairies, Oklahoma and Nebraska	446	1.22 (3.0)
Short grass grassland, 13 in. rainfall; Wyoming	69	0.19 (0.5)
Desert, 5 inches rainfall, Nevada	40	0.11 (0.2)
Seaweed beds, Nova Scotia	358	1.98 (1.0)

* Values are grams of dry organic matter (from Odum, 1959). Values in parenthesis are rates for growing season only, which is often less than a year.

than sunlight (as in the lush forest) or water (as in the desert) often limits the amount of algal production.

There is good evidence that the production of plant material has reached some sort of optimum level. That is, production depends mostly on climate and soil, and it appears in most regions to have reached the level at which climate and soil alone are critical. This means that a region is occupied by those plant species that are best suited to produce vegetation in the climate of that region, for otherwise knowledge of plant species as well as climate would be essential to prediction of productivity. Fig. 7-3, drawn from data gathered by Rosenzweig, shows productivity and evapotranspiration (which is the excess of yearly precipitation over runoff, and, therefore, measures the amount of water evaporated or transpired). From this figure it is clear

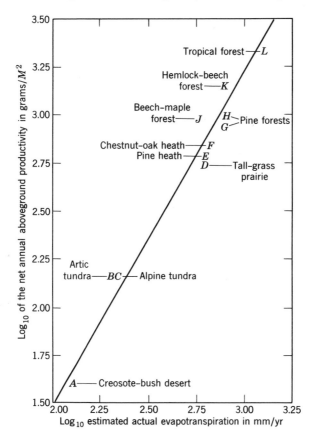

Fig. 7-3. The relation between production and a climate measure. See text for further discussion. (The graph and the idea were supplied by M. Rosenzweig.)

that from a knowledge of evapotranspiration we can make a good guess about production of natural vegetation. Because evapotranspiration is great when both rainfall and temperature (which speeds up evaporation and transpiration) are great, tropical areas can be expected to show greater production.

We have seen that some of the sun's energy is stored in green plants, which are the source of most of the energy utilized by all animals. In fact, the herbivorous animals, whether they are grazing cows or "grazing" minute crustacea that eat the floating algae, can all be lumped together as organisms which derive their energy and nutrients directly from green plants.

We all know that the amount of plant material influences the amount of animal life in some way. We are familiar at least with the results of temporary plant food shortages for human populations—the Irish potato famine and the perpetual hunger in modern India and China, for example. On the other hand, it is harder at present to associate population increases with temporary surpluses of plant food. However, people are perhaps exceptional in this respect: the human population has been growing without interruption for thousands of years (see Fig. 5-4). Even the "black death" and the worst wars and famines have been insignificant in the general trend of increase in the world's population. Therefore, we are not in a position to say what will limit human populations; man is quick to design new tools to cope with new emergencies. Most other species are different in that they are unable to increase their food supply by cultivating new areas. For these species it is obviously important to make effective use of available food.

If we wish to investigate how many units of plant are required to produce a unit of herbivore, we must choose some unit in terms of which we can measure both the plants and the animals. We could count individuals of both, for example. On reflection, however, we may prefer some unit that did not equate a single grass plant with a single tree. Perhaps we should use volume, but it too is inappropriate unless we know how compressed each unit of volume is. Weight would be better, because a gram of grass seems roughly equivalent to a gram of tree as food for animals. Even better, however, is to measure the fuel value of each, because much of our food is eaten for its energy. A convenient unit of fuel value, or energy, is the calorie. From people who are concerned with gaining or losing weight, we learn that a gram of walnut meats contains 7.28 kilogram calories (i.e., it contains enough energy to heat a kilogram of water 7.28 Centigrade degrees), and a gram of another plant food, lettuce, contains only 0.19 kilogram calories. In other words, a gram of walnuts contains 38 times as much fuel value as a gram of lettuce. Although not all of this energy may be available to the animal, calories still provide the best unit of comparison.

What is the fate of a calorie of food energy that enters an animal's mouth? It is either used, wasted, or stored. The second law of thermodynamics says that no process is fully efficient, or, in other words, that a positive fraction of the energy is always wasted. A sizable amount is also used up just to maintain the animal—either the energy used in moving about or that which does the work of building up and breaking down body proteins, etc. Although these used-up calories are of the greatest interest to the biochemist and are measured by the amount of respiration, a carnivore is primarily concerned with the amount stored as new matter in his prey. For the carnivore—a cat, for example—looks at an herbivore—a mouse, for example—as a factory for converting plant energy to some form of energy palatable to a carnivore. In the long run, just as a prudent investor lives from the interest of his investments and leaves the principal invested to gather new interest, so the prudent cat must always leave a healthy mouse population as "principal" from which to reap the "interest" for his food. This interest is the excess of the mouse population of one generation over that of the previous generation, and therefore represents the excess of the amount of food energy which was stored (perhaps in new individuals) over that used or wasted. (As we noted in Chapter 1, the interest is also the fitness, so that we have, incidentally, a connection between fitness and the energy relationships now under discussion).

We can now define the "ecological efficiency" of the mouse (as an agent for converting plants into cat food) as the ratio

$$\frac{\text{calories of mice eaten by cats per unit time}}{\text{calories of plants eaten by mice per unit time}}$$

Using *Chlamydamonas*, an alga, as the plant food, *Daphnia* the "water flea" as the herbivore, and investigators from their own laboratory as the carnivores, Slobodkin (1960) with his students was able to measure ecological efficiency of *Daphnia* with considerable accuracy. Richman (1958), instead of eating the *Daphnia* that he harvested as a true carnivore would do, burned them in a calorimeter to measure their energy content; he also burned samples of *Chlamydamonas* and so was in a position to give accurate estimates of the numbers of calories both eaten by the *Daphnia* and harvested by the "carnivore." He always found the maximum efficiency to be about 10 per cent, that is, *Daphnia* can convert about 10 per cent of the plant food that they digest into increased *Daphnia* population, which can be harvested without harming the future production of new *Daphnia*. Slobodkin also summarizes many less precise measurements that have been made in natural populations with real carnivores, who eat their harvest of herbivores. Most of these also show an ecological efficiency of about 10 per cent. This is remarkable, because it looks as if carnivores really do harvest their herbivore prey at the maximum prudent rate, in spite of the fact that we might expect the selfish process of natural selection to lead to overexploitation of prey with resulting reduction of yield. This is remarkable also because the estimates are quite uniform, suggesting that ecological efficiency is about 10 per cent, whether the herbivore is a *Daphnia* or a cow.

H. T. Odum (1957) studied the amounts of energy per year entering plants (as sunlight), entering herbivores (as their plant food), entering carnivores (as their herbivore food), and entering carnivores that eat carnivores (as their carnivore food) in Silver Springs, Florida. Although field data of this kind are less precise than the laboratory work mentioned earlier, they give the following picture (numbers are of kilogram calories under each square meter of water surface per year) of the food chain:

Plants		Herbivores		Carnivores		Secondary carnivores	
20810	8833	3368	1478	383	67	21	6
entering	stored	entering	stored	entering	stored	entering	stored

The ecological efficiencies are: plants $3368/20810 = 16$ per cent; herbivores $383/3368 = 11$ per cent; carnivores $21/383 = 5.5$ per cent. Notice that these ecological efficiencies are in the neighborhood of 10 percent as Slobodkin noted. We can use this to make an interesting calculation. There are four stages in this food chain: plants, herbivores, carnivores, and secondary carnivores. Each stage can expect to take in energy at about 1/10 the rate of the previous stage; hence the secondary carnivores are reduced to roughly $1/10 \times 1/10 \times 1/10 = 1/1000$ of the energy taken in by the plants. Supercarnivores which ate these would be reduced to 1/10th of this or 1/10,000 of what the plants received, and so on. No wonder that very few species find it worthwhile to be such a supercarnivore—there is practically no energy available to them!

One of the longest natural food chains that we have thought of begins with single-celled plants in ocean plankton; these are eaten by small crustaceans, which are eaten by small fish, which are eaten by large fish, seals, or porpoises, which are then eaten by killer whales.

Whether a longer food chain than this one of five links can exist we do not know. We can see why natural selection might favor a shortening of food chains. Some of the largest animals, the baleen whales, have shortened their chain to three links: planktonic plants—crustaceans—whales (Fig. 7-4). The same, of course, also applies to humans; if they act as herbivores—most races on the verge of starvation do—they can rely on ten times as much available food energy on the earth as they can if they act as pure carnivores, as, for example, the Eskimos do.

Another feature of this 10 per cent figure is that it seems to vary little with the number of species in the community being studied. Slobodkin's laboratory system had one species of herbivore; Odum's and other field studies have had from a few to many hundreds of species. Nevertheless, all have about 10 per cent ecological efficiency. In other words, the number of species has little to do with the overall efficiencies of plants, herbivores, carnivores, etc.

We shall now relate the number of individuals actually present at a given time to the energy flow terms we have just been using. Suppose, for example, that the carnivore population (cats) are eating 10,000 kilogram calories of herbivore (mice) per day. If each mouse weighs 20 grams and each gram of mouse has 5 kilogram calories, each mouse represents 100 kilogram calories, so that the cat population must be eating 100 mice per day. We can imagine two different regimes of mouse production, that would account for this. As one extreme, suppose that each mouse lived thirty days, after which it was eaten by a cat. Then 1/30 of the mice are being eaten every day. But this 1/30 must supply 100 mice per day to the cats, so that there must be 3000 mice in our population. In other words, a standing crop of 3000 mice with a mean life-span of 30 days will provide the necessary 100 mice per day. Now suppose instead that each mouse lived 300 days, so that only 1/300 of the mice were eaten each day. Then a standing crop of 30,000 would be required, so that the cats could get 100 per day. In general terms

$$\frac{\text{"standing crop" in number of organisms}}{\text{mean life-span in days}} = \frac{\text{number of organisms}}{\text{dying/day}}$$

$$\frac{\text{standing crop in grams}}{\text{mean life-span in days}} = \text{number of grams dying/day}$$

$$\frac{\text{standing crop in calories}}{\text{mean life-span in days}} = \text{number of calories dying/day}$$

When we solve it, standing crop in calories = mean life-span (days) × calories dying/day. Notice further that, because the amounts of energy

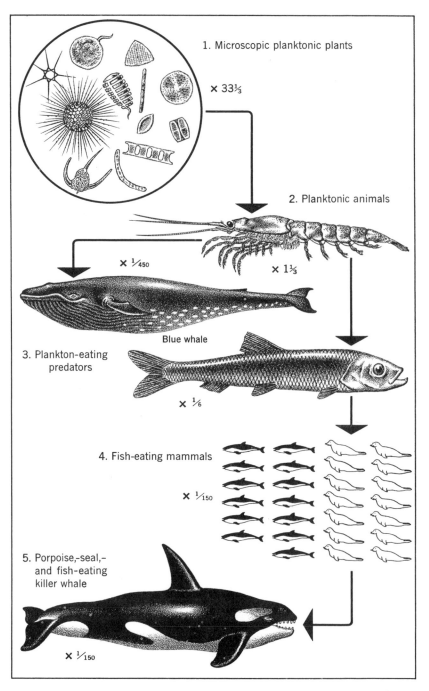

1. Microscopic planktonic plants

× 33⅓

2. Planktonic animals

× 1/450

× 1⅓

Blue whale

3. Plankton-eating predators

× 1/6

4. Fish-eating mammals

× 1/150

5. Porpoise,-seal,- and fish-eating killer whale

× 1/150

Fig. 7-4. Two food chains of different lengths. One killer whale's stomach contained thirteen porpoises and fourteen seals.

flowing through links in the food chain remains fairly constant from day to day, the calories of animals dying must be exactly replaced by new calories of stored energy.

In any case, if we know the kind of organisms that we are dealing with and their mean life-span (or, more accurately, the mean life-span of any given gram of the organism), we can convert the energy pattern, which we know, to a standing crop, and if we know the mean weight of individuals, we can say how many individuals the population will contain.

Because mean life-span of plankton organisms is short—a matter of days—the standing crop is small. On the other hand, the same number of calories in a terrestrial community with long-lived herbivores (e.g., deer) might require a very large standing crop.

We pointed out in Chapter 5 that small organisms have shorter life-spans than large ones. Therefore, from the previous paragraph, we know that when prey species are small, a smaller standing crop is required to support a given predator population. This creates some seemingly paradoxical situations. For example, under a square meter of surface water in Long Island Sound, 16 grams of tiny planktonic plants will support 32 grams of herbivores feeding on them; in Silver Springs, where the plants were large, 809 grams supported 37 grams of herbivores (Odum, 1959). The paradox vanishes when we remember* that smaller organisms metabolize and grow faster in proportion to their weight than large ones. The tiny planktonic plants may double their weight every day, whereas larger aquatic plants may take several weeks to do so. Small organisms, threatened by more dangers from the environment, live shorter lives and must grow and reproduce more quickly to maintain their populations.

THE NUMBER OF SPECIES

Next we inquire among how many species the individuals are divided. Are they all of the same species, or do they belong to many? The basic principle governing our answer is that "a jack-of-all-trades is a master of none" in other organisms as well as humans. Hence a single species utilizing the whole spectrum of resources (which is therefore a jack-of-all-trades and hence a master of none) will be outdone in competition by a combination of specialists, each geared to utilize a particular range of resources with maximum efficiency. This is why all

---

* See W. Telfer and D. Kennedy, *The Biology of Organisms*, Wiley, 1965, pages 201–204.

of the carnivores in the world are not of the same generalized species. In other words, we cannot picture a single carnivore which is simultaneously as good as a baleen whale at catching shrimps, as good as a lion at catching antelopes, and as good as a predatory wasp at catching caterpillars. However, we must not carry this argument too far, for it will lead to the conclusion that every individual should belong to its own unique species, which would give us as many species as there are individuals! This is, of course, preposterous, for sexual reproduction with its benefits would then be impossible. We therefore expect the number of species to reach some intermediate level which is sufficiently large to permit fairly effective specialization but not so large that the species become precariously rare.

Before we proceed further in studying the number of species in an environment, it is useful to draw an analogy with the number of books in a library. If a library is known to be full with no more room on its shelves, the number of its books can be predicted from a knowledge of the length of the shelves and the distance between centers of adjacent books. If, on the other hand, the library is not full, the number of books can only be predicted from knowledge of the rate of acquisition (and elimination) of books and the age of the library. Thus it is vital to the prediction of the number of books in the library that we know whether or not it is full. Similarly, to understand the number of species in an environment, we first must decide whether the environment is "full" of species, or, in the usual terminology, whether the environment is saturated with species. In our study of competition we saw that there is likely to be a limiting similarity of coexisting species (analogous to the distance between centers of packed books), so that saturation is indeed possible. Whether saturation has been reached is another matter, and biologists argue heatedly about this without generating much light.

There are many patterns of species diversity that need explaining. Suppose, for example, that we count the number of bird species which breed in the following areas from arctic regions to the tropics. We obtain: Alaska, 222; British Columbia, 276; Washington, 235; Oregon, 232; California, 286; Mexico, 764; Guatemala, 472; Nicaragua, 477; Costa Rica, 603; Panama, 667. Considering that the Central American countries are smaller than Alaska or British Columbia, there is an astonishing increase in number of species as we enter the tropics. This tropical increase is not confined to birds; it is well-nigh universal, if we consider large groups of plants and animals. When applied to smaller groups, such as salamanders or coniferous trees, there are numerous exceptions.

Another pattern of species diversity is evident on comparing islands with the mainland. Small and remote islands have fewer species than large islands nearer the source of colonization. Very remote islands, such as Easter Island in the Pacific, may have no resident reptiles or mammals and few insects. The library analogy is appropriate again: are there fewer species on remote islands because fewer have ever reached there, or because remote islands will maintain fewer? The first is obviously a possibility, but the second may be just as likely. In Fig. 7-5 we plot the rates of immigration of new species (not on the island) and of extinction of species from the island against the number of species on the island. When the island has few species on it, most of the immigrants are of new species, so that the immigration curve is high. When the island already contains most of the colonizing species, few of the immigrants will be new, so that the immigration curves are lower on the right. Similarly, the more species there are on the island, the more can go extinct, so that the extinction curves rise. Hence the extinction curves must intersect the immigration curves. Where they intersect, the rate of extinction just balances the rate of

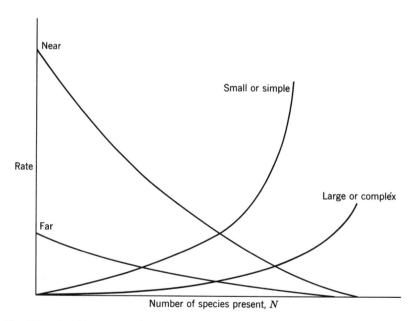

Fig. 7-5. Equilibrium model of faunas of several islands of varying distances from the source area and of varying size. Descending curves are rates of immigration of new species; ascending curves are rates of extinction. Where they intersect, there is a stable equilibrium. (After MacArthur and Wilson, 1963.)

immigration. This is the equilibrium number of species (MacArthur and Wilson, 1963). Because remote islands have lower immigration rates, they would have fewer species at equilibrium when small or simple. Hence either the historical (unsaturated) or equilibrium (saturated) explanations could account for the main island patterns. Again, there is relatively little evidence of which explanation is correct for any given group of organisms. Perhaps the best evidence comes from the histories of islands such as Krakatau (see page 51 for a fuller account), which reconstituted a bird fauna about equal to that of other islands of comparable size and remoteness within fifty years after it was devastated by the volcanic explosion of 1883. This strongly suggests the correctness of the equilibrium explanation for birds, at least for islands near the source of colonization. It is less certain for other organisms, which take longer to colonize. Doubtless many forms on many islands are unsaturated.

One final pattern of species diversity is illuminating. We begin again with the library analogy. Suppose that a given library with limited shelf space decides to shelve only one book by each author or only one book on each of a variety of subjects. The first single book that is acquired by this library will, of course, be shelved because it will be the only one of its kind. The second book given to it is likely to be different from the first and it, too, will have an appropriate shelf space. As the library randomly accumulates more books, the chance of acquiring a second book on the same subject increases, and eventually such books are certain to arrive. These will be sent on to neighbor libraries, in case they still need a book of this type. Eventually the library will be full, and, although substitutions of a better book may be made for a less good alternative, the library size will not increase. All extra books will at this point be sent to other libraries.

Small and comparatively uniform pieces of environment (which we usually call habitats) are like these hypothetical libraries; they also have a limited capacity for species. The first few species that colonize are almost certain to be successful in all habitats for which they are suited. Eventually, however, each habitat is virtually saturated, and additional immigrant species cannot increase the number of species per habitat, although they may replace one. They can increase the number of species in the whole area, comprising several habitats, however, for they may replace an existing species in one habitat, but not replace the same species in a second habitat. Thus later immigrant species may increase the total number of species in an area having several different habitats but will not increase the number of species per habitat. This phenomenon is easiest to witness on islands, which usually have a small sample of the mainland species. The island species then occupy a wider variety of habitats than they did on the mainland. (MacArthur, 1965) An additional species arriving from the mainland might restrict the habitats of the others, but does not necessarily thereby increase the number of species per habitat.

APPENDIX

Suggestions for Laboratory Exercises*

We have pointed out that a great length of time is required for most evolutionary changes. Therefore, laboratory exercises using real organisms cannot possibly follow a complete history of the change. There are two alternatives: we can study, in nature, the *outcomes* of evolutionary changes, or we can study, with objects that reproduce much faster than real organisms, the *mechanisms* of evolutionary change. Here we outline some laboratory exercises of the latter type, designed to accompany the chapters in this book. Wherever possible, they should be supplemented by field trips in which real organisms rather than imitations are studied.

The materials for this laboratory are beads, lead shot, or some other objects that can be obtained in two or three colors but are otherwise identical. Students should work in groups of two or three; each group should have about 1000 beads, of two colors—say, half of them red and half black. Each group should also have a long drawer or box, along the bottom edge of which the beads can be lined up. A good deal of grade school arithmetic is involved in the calculations, and cheap slide rules would speed up this aspect of the work considerably, but are not essential. Pencils (preferably of two colors), paper, and graph paper are the only other requirements.

---

* The idea for these exercises came from M. Williamson and P. Sheppard.

FITNESS

Here each bead is to represent an organism, or, equally well, a gene. We let the beads reproduce and die, and compute their fitness. Put 100 beads of any color into the drawer (or tray or box); it represents the population of one generation. Suppose that the fitness of the beads is 10 per cent. That is, they accumulate at a rate of 10 per cent per generation. Then, in the next generation 10 per cent of 100 which is 10 new individuals will be present. Therefore, add 10 news beads to the drawer, making a total of 110. The next generation 10 per cent of 110 which is 11 new beads will be added, making 121. The next generation 10 per cent of 121 which is 12.1 should be added. This is nearer to 12 than 13, so add 12 beads. Continue in this way for five generations and plot the results on a graph in which the generations are plotted on one axis and the number of organisms (beads) along the other. (In this simple example you do not really need the beads; you can calculate with pencil and paper alone. Later the beads are more essential.)

We now make the process more realistic by letting the beads give birth and have deaths. First, let each of 10 beads give birth to two new ones apiece, by adding 10 new beads to the drawer. Next, one half of the class should let precisely half of all of the beads die by removing half of them (for concreteness, when taking half of an odd number, follow the procedure of taking just too many the first time and just too few the next). The other half of the class will kill about half, too, but in the following way: flip a coin for each bead and remove it if the coin comes up heads (any other device for producing random mortality of the probability of one half is equally good). Because both halves of the class *expect* the same mortality rates, the fitness is the same in the populations investigated by both halves of the class. What is this fitness? (That is, by what per cent do we expect the next generation's population to be greater?) Does the population always grow at precisely the rate given by the fitness? At what rate could a predator remove individuals from this population if it wanted to leave a constant number of individuals after predation? The student can act as a predator and remove to a different box the predated beads, taking just enough to leave the same number of beads in each generation. Notice that the per cent of the population that is taken by the predator is equal to the fitness. This, then, is a new definition of fitness. Write the definition out carefully.

NATURAL SELECTION
Part A

Here we let beads represent genes. Each black bead is one gene and each red bead is a different (allelic) gene. Begin with a population of gametes containing 100 red and 100 black genes. Let them "mate randomly" by shaking them up together in the drawer (this is the random part) and then, by tipping the drawer, tap them into line along a bottom edge of the drawer. When they are arrayed in a single file in random order, mark them into pairs from one end; this is the "mating." Suppose, for example, that the order of the beads in the line is $R, R, B, R, R, B, R, R, \ldots$, etc. We mark them into pairs, each pair representing an diploid zygote, in this way: $(R, R), (B, R), (R, B), (R, R), \ldots$, etc. Each pair represents the genotype of an individual produced by random mating. (R, B) and (B, R) are both normal heterozygotes and cannot be distinguished (biologically, of course).

Now let us suppose that red is recessive and, in homozygous form, lethal. This means that all (R, R) pairs are killed, which in our case means that they are removed from the drawer. About 50 red beads will usually have been removed; this alters the proportion of red and black beads and also reduces the total number of beads. Plot not only the total number of red and black beads after each generation of selection, but also, on a different graph, the proportion of beads that are red. After removing the beads form the next generation by again considering the remaining beads gametes, and thus again produce a random mating. Continue this for five generations or until the red genes disappear. How does this process fail to duplicate natural selection against a lethal recessive? Can you redesign the experiment to remedy these defects? Repeat the process with the dominant gene assumed to be lethal; compare the speeds of selection.

Part B

We now make the process more realistic. One part of the class is to use the following fitnesses: (RR) -40 per cent (i.e., 40 per cent of (RR) individuals will be removed); (RB) -10 per cent; (BB) 0 per cent (i.e., none will be removed or added). Carry this selection out for five generations, plotting your results. Begin with 100 of each color.

The other part of the class is to verify the situation of balanced polymorphism by using the following fitnesses: (RR) -20 per cent; (RB) 0 per cent; (BB) -40 per cent. When the total population gets below 50, reconstitute it—by maintaining the same proportion of red and black—to a larger size. Plot the proportion of red genes for ten generations. Why does neither gene disappear? Did anyone find one gene disappearing? Why? Would this have happened in a larger population? Compare results with the other half of the class.

Part C: Fundamental Theorem of Natural Selection

Fisher has proved a theorem about the rate of increase of the average fitness of a population. The technical details of this are beyond the scope of this book, but the essence of it can be verified readily with the beads. Let each group of students design its own selection experiment with two alleles (or, if three or more colors of bead are available, with three or more alleles). Assign fitnesses to each genotype (keep them negative or zero to avoid having to decide what the offspring genotypes will be; if the populations get too small, reconstitute them to a larger size as before). Plot the average fitness of all of the individuals for several generations of selection. (For example, if the frequencies and fitnesses of genotypes are

genotype	RR	RB	BB	
fitness	−10%	−5%	0%	
number of individuals	37	17	23	Total: 77

then the average fitness of this population is given by

$$\frac{37(-0.10) + 17(-0.05) + 23(0)}{77} = \frac{-4.55.}{77} = -0.059)$$

Does the average fitness ever decrease? Would it in a large population? Does it always increase at the same rate? In which groups did the fitness increase most rapidly? (A complete characterization of the rate of increase in fitness requires the tools of statistics.) We have held the fitnesses of the genotypes constant; if we let them change with varying environmental conditions, will fitness still never decrease (except by chance)?

Part D: The Cost of Natural Selection

Our previous models have been unrealistic in at least one way which we now correct. Real populations are usually maintained at a fairly constant level. Our population decreased because the mean fitness was always negative. To correct this, we maintain the population at 200 beads in the following way. If, after selection, we have 75 black beads and 50 reds, making a total of 125 (of which 50/125 or 40 per cent are red), we begin the next generation not with these beads, but with 80 reds and 120 blacks, so that still 40 per cent are red, but the total is 200. With these beads we perform the random mating that produces the gametes which we select, and so on. This correction does not affect the change in proportion of the red beads with time, for this was determined by the relative fitnesses of the genotypes that are still unchanged. Hence our previous results would be unchanged by this correction. However, it is now easier to demonstrate Haldane's useful results. (The s of page 94 is now 1.)

Divide the class into two parts. Each part begins with 20 red beads and 180 black ones and continues selecting, as directed below, for the fitter red ones until there are 20 or fewer blacks and at least 180 reds. Both parts of the class will have heterozygotes precisely intermediate in fitness between the homozygotes (any other uniform degree of dominance would do). One part will assume the per cent removed to be: (RR) 0 per cent; (RB) 20 per cent; (BB) 40 per cent, whereas the other part of the class will remove the following percentages: (RR) 0 per cent; (RB) 10 per cent; (BB) 20 per cent. Both parts of the class should plot the progress of selection (proportion red), and both should keep track of the number of beads *removed* each generation. Each *pair* of beads removed is a selective death. (Why? Would it still be if we also removed some of the (RR) genotypes?) Each group should calculate the total number of selective deaths occurring in the substitution, and they should be averaged in each part of the class. How many selective deaths occurred, per individual? (There were always 100 individuals.) Is the average number of selective deaths about independent of the strength of selection? Yet the number of generations required for the substitution was very different in the two parts of the class. How then can you account for the equal number of selective deaths?

PREDATOR-PREY INTERACTIONS

In this experiment, red beads represent predators and black beads their prey. No genes are involved and we need not worry about random matings. However, capture of a prey is a chance affair: the predator in his wanderings must happen on a prey individual. We imitate this process by shaking the beads, lining them up randomly in single file, and pairing them off from one end as before. When a pair contains a red and a black bead, we will say that the predator (red bead) has captured the prey (black bead). Because he will eat it, and then be able to reproduce, we will remove the black bead from each such pair and replace it with a red one (or with 2 red ones if we wish to assign a higher birthrate to the predator). Red beads not paired with black ones are predators who came upon no prey and hence starved to death; remove these beads! Black beads paired with other blacks are prey who were not eaten and hence were able to reproduce. To each such bead add one (or 2, or more for higher birthrates of prey) to represent new prey individuals.

Divide the class into three parts. Each part will carry through 10 generations and each will begin with 100 red and 100 black beads.

Part 1: Use prey birthrate 1 and predator birthrate 2.

Part 2: Use prey birthrate 2 and predator birthrate 1.

Part 3: Use prey birthrate 2 and predator birthrate 1, but assume that the prey population cannot exceed 300 (because of insufficient food or hiding place). Hence, this part of the class will remove any excess of prey over 300 immediately following reproduction.

Plot results, generation by generation. By comparing results of all parts of the class, answer the following questions. Can predator exterminate prey? When? Can prey increase so that predators never catch up? When? Can oscillations be set up? When? Are these only random, or would they persist even in a very large population?

FURTHER EXERCISES

A class that has completed these exercises can design any number of others that are very instructive. In fact, designing such exercises helps more to solidify one's knowledge of population biology than does carrying out the instructions!

The student who can design a bead experiment to demonstrate selection against a sex-linked gene or selection when there is inbreeding will really understand these processes. (Inbreeding, to a given degree —say 40 per cent—means that 40 per cent of the genes combine with their own kind to form homozygotes, whereas the remaining 60 per cent mate randomly.) Other interesting topics that can be clarified with the beads are the effect of subdividing a population into parts that always mate within the parts, and genetic drift (random changes in gene frequency, caused by the accidents of sampling and mating.

The student has seen many of them but has not analyzed exactly when they are most likely to be important). If more patterns of beads are available, selection at two or more loci (linked or unlinked) simultaneously is a rewarding experiment.

People who perform "computer simulations" of genetic or ecological situations are often simply letting the computer run through many generations of bead experiments.

Bibliography[*]

Allee, W. C. (1951). *Cooperation among animals.* Schuman, New York.

Allison, A. C. (1956). The sickle cell and hemoglobin C genes in some African populations. *Ann. Hum. Genet.* **21**:67–89.

Andrewartha, H. G., and L. C. Birch (1954). *The distribution and abundance of animals.* Univ. of Chicago Press.

Baker, F. S. (1950). *Principles of silviculture.* McGraw-Hill, New York.

Brower, J. V. Z. (1958). Experimental studies of mimicry in some North American butterflies. Part I. *Evolution* **12**:32–47.

Brower, J. V. Z. (1960). Experimental studies of mimicry. IV. The reactions of starlings to different proportions of models and mimics. *Am. Naturalist* **94**: 271–282.

Brown, W. L., Jr., and E. O. Wilson. (1956). Character displacement. *Syst. Zool.* **5**:49–64.

Byers, H. G. (1954). The atmosphere up to 30 kilometers. In *The earth as a planet,* G. P. Kuiper, Ed. Univ. of Chicago Press.

Cain, S. (1935). Studies of virgin hardwood forest: III. Warren's woods, a beech-maple climax forest in Berrien County, Michigan. *Ecology* **16**:500–513.

Christian, J. (1959). The roles of endocrine and behavioral factors in the growth of mammalian populations. In *Symposium on comparative endocrinology,* A. Gorbman, Ed. Wiley, New York.

Clark, W. E. LeGros (1963). *The antecedents of man.* Harper Torchbooks, New York. (First published in 1959 by Edinburgh University Press, Edinburgh, Scotland.)

Clausen, J. (1951). *Stages in the evolution of plant species.* Cornell Univ. Press., Ithaca, New York.

Cole, L. (1954). The population consequences of life history phenomena. *Quart. Rev. Biol.* **29**:103–137.

Connell, J. (1961*a*). Effects of competition, predation by *Thais lapillus* and other factors on natural populations of the barnacle *Balanus balanoides. Ecol. Monographs* **31**:61–104.

Connell, J. (1961*b*). The influence of interspecific competition and other factors on the distribution of the barnacle *Chthamalus stellatus. Ecology* **42**:710–723.

Connell, J. (1963). Territorial behavior and dispersion in some marine invertebrates. *Res. Pop. Ecol.* **5**:87–101.

Connell, J. (1964). Studies on predator-prey interactions in intertidal animals at Friday Harbor, Washington. Unpublished report.

---

[*] Journal abbreviations are those adopted by the American Chemical Society as an international standard. *See Chemical Abstracts List of Periodicals.*

Connell, J., J. G. Tracey, and L. J. Webb. (1964) Studies of species diversity of rainforests in Queensland, Australia. Unpublished report.

Crombie, A. C. (1946). Further experiments on insect competition. *Proc. Roy. Soc. London (B)* **133**:76–109.

Crombie, A. C. (1947). Interspecific competition. *J. Animal Ecol.* **16**:44–73.

Dammerman, K. W., (1948). The fauna of Krakatau 1883–1933. *K. Nederland Akad. Wetensch. Afd. Natuurkunde sect* **2**:1–594.

Darlington, P. J. (1957). *Zoogeography.* Wiley, New York.

Darlington, P. J. (1965). *Biogeography of the southern end of the world.* Harvard Univ. Press, Cambridge, Mass.

Darwin, C. (1859). *On the origin of species.* Murray, London.

Darwin, C. (1860). *Journal of researches—during the voyage of H. M. S. Beagle.* Various publishers. This quote is from the 1906 edition published by Dutton, New York.

Darwin, C. (1959). *Autobiography,* Nora Barlow, Ed. Harcourt, Brace and Co., New York.

Darwin, C., and A. R. Wallace. (1958). *Evolution by natural selection.* Cambridge Univ. Press.

Deevey, E. S. (1947). Life tables for natural populations of animals. *Quart. Rev. Biol.* **22**:283–314.

Deevey, E. S., Jr., (1960) The Human Populations. © Scientific American Inc., 203 (Sept.): 194-204. All rights reserved.

Dodd, A. P. (1959). The biological control of prickly pear in Australia. In *Biogeography and ecology in Australia,* A. Keast, R. L. Crocker, and C. S. Christian, Eds. Dr. W. Junk. The Hague.

Ehrlich, P. R., and R. W. Holm (1963). *The process of evolution.* McGraw-Hill, New York.

Elton, C. S. (1958). *The ecology of invasions by animals and plants.* Methuen Monograph, Wiley, New York.

Etkin, W., Ed. (1964). *Social behavior and organisation among vertebrates.* Univ. of Chicago Press.

Fisher, J., and R. T. Peterson. (1964). *The world of birds.* Doubleday, New York.

Fisher, R. A. (1958). *The genetical theory of natural selection.* Dover, New York.

Flint, R. F. (1957). *Glacial and pleistocene geology.* Wiley, New York.

Frank, P. (1952). A laboratory study of intraspecies and interspecies competition in *Daphnia pulicaria* (Forbes) and *Simocephalus vetulus* (O. F. Muller). *Physiol. Zool.* **25**:173–204.

Frank, P. (1957). Coactions in laboratory populations of two species of *Daphnia.* *Ecology* **38**:510–519.

Gause, G. F. (1934). *The struggle for existence.* Williams and Wilkins, Baltimore. (Reprinted 1965 by Hafner Publishing Co., New York.)

Guhl, A. M. (1953). *Social behavior in the domestic fowl.* Tech. Bull. No. 73, Agr. Exp. Sta., Kansas State College, Manhattan, Kansas.

Haldane, J. B. S. (1957). The cost of natural selection. *J. Genet.* **55**:511–524.

Harrison, J. L. (1962). The distribution of feeding habits among animals in a tropical rain forest. *J. Animal Ecol.* **31**:53–63.

Hatton, H. (1938). Essais de bionomie explicative sur quelques espèce intercotidal d'algues et d'animaux. *Ann. Inst. Oceanogr. Monaco.* **17**:241–248.

Hediger, H. (1955). *Psychology of animals in zoos and circuses.* Criterion Books, New York.

Holdridge, L. R. (1947). Determination of world plant formations from simple climatic data. *Science* **105**:367–368.

Huffaker, C. B. (1958). Experimental studies on predation. *Hilgardia* **27**:343–383.

Hutchinson, G. E. (1953). *The itinerant ivory tower.* Yale Univ. Press., New Haven, Conn.

Hutchinson, G. E. (1965). *The ecological theater and the evolutionary play.* Yale Univ. Press., New Haven, Conn.

Karn, M. N., and L. S. Penrose (1951). Birth weight and gestation time in relation to maternal age, parity and infant survival. *Ann. Eugen.* **16**:147–164.

Kettlewell, H. B. D. (1956). Further selection experiments on industrial melanism in the Lepidoptera. *Heredity* **10**:287–301.

Klopfer, P. (1962). *Behavioral aspects of ecology.* Prentice-Hall, Englewood Cliffs, N.J.

Koeppe, C. E. (1958). *Weather and climate.* McGraw-Hill, New York.

Lack, D. (1947). *Darwin's finches.* Cambridge Univ. Press.

Lack, D. (1954). *The natural regulation of animal numbers.* Oxford Univ. Press.

Lack, D. (1957). *Evolutionary theory and Christian belief.* Methuen, London.

Lea, D. E., and C. A. Coulson (1949). The distribution of the numbers of mutants in bacterial populations. *J. Genetics* **49**:264–285.

Lederberg, J., and E. Lederberg (1952). Replica plating and indirect selection of bacterial mutants. *J. Bacteriol.* **63**:399–406.

Levins, R. (1962). Theory of fitness in a heterogeneous environment. *Am. Naturalist* **96**:361–373.

Lloyd, M. (1966). Mean crowding. In press.

Luria, S., and M. Delbrück (1943). Mutations of bacteria from virus sensitivity to virus resistance. *Genetics* **28**:491–511.

Macan, T. T. (1963). *Freshwater ecology.* Wiley, New York.

MacArthur, R. (1962). Some generalized theorems of natural selection. *Proc. Nat. Acad. Sci.* **48**:1893–1897.

MacArthur, R. (1965). Patterns of species diversity. Biol. Rev. **40**:510–533.

MacArthur, R., and R. Levins (1964). Competition, habitat selection and character displacement in a patchy environment. *Proc. Nat. Acad. Sci.* **51**:1207–1210.

MacArthur, R., and E. O. Wilson (1963). An equilibrium theory of insular zoogeography. *Evolution* **17**:373–387.

Marler, P. (1959). Developments in the study of animal communication. In *Darwin's Biological Work,* P. R. Bell, Ed., Cambridge Univ. Press.

Mayr, E. (1963). *Animal species and evolution.* Belknap Press, Cambridge.

Merriam, D. (1941). Studies of the striped bass (*Roccus saxatilis*) of the Atlantic coast. *Fish Bull., U.S. Fish and Wildlife Service* **50**:1–77.

Mooney, H. A., and W. D. Billings (1961). Comparative physiological ecology of arctic and alpine populations of *Oxyria digyna. Ecol. Monographs* **31**:1–29.

Newell, N. D. (1963) Crises in The History of Life. © Scientific American Inc.,208 (Feb.): 76-92. All rights reserved.

Odum, E. P. (1959). *Fundamentals of ecology.* Saunders, Philadelphia.

Odum, E. P. (1963). *Ecology.* Holt, Rinehart and Winston, New York.

Odum, H. T. (1957). Trophic structure and productivity of silver springs, Florida. *Ecol. Monographs* **27**:55–112.

Organ, J. A. (1961). Studies of the population dynamics of the salamander genus *Desmognathus* in Virginia. *Ecol. Monographs* **31**:189–220.

Osborn, R. H., and F. V. De George (1959). Genetic basis of morphological variation. Harvard Univ. Press, Cambridge, Mass.

Ovington, J. D. (1962). Quantitative ecology and the woodland ecosystem concept. *Adv. Ecol. Research* **1**:103–192.

Paine, R. T. (1966). Food web complexity and species diversity. *Amer. Naturalist* **100**:65–75.

Paris, O. H., and F. A. Pitelka (1962). Population characteristics of the terrestrial isopod *Armadillidium vulgare* in California grassland. *Ecology* **43**:229–248.

Park, T. (1948). Experimental studies of interspecific competition. I. Competition between populations of the flour beetles *Tribolium confusum* Duval and *Tribolium castaneum* Herbst. *Ecol. Monographs* **18**:265–308.

Park, T. (1954). Experimental studies of interspecific competition. II. Temperature, humidity and competition in two species of *Tribolium*. *Physiol. Zool.* **27**:177–238.

Pitelka, F. A., P. Q. Tomich, and G. W. Treichel (1955). Ecological relations of jaegers and owls as lemming predators near Barrow, Alaska. *Ecol. Monographs.* **25**:85–117.

Richman, S. (1958). The transformation of energy by *Daphnia pulex*. *Ecol. Monographs* **28**:273–291.

Ricker, W. E. (1954). Compensatory mortality. *J. Wildlife Management* **18**:45–51.

Romer, A. S. (1945). *Vertebrate Paleontology*. Univ. of Chicago Press.

Rosenzweig, M., and R. MacArthur (1963). Graphical representation and stability conditions of predator-prey interactions. *Am. Naturalist* **97**:209–223.

Salisbury, E. J. (1925). The structure of woodlands. In *Festschrift Carl Schröter*. *Veroff. geobot. Inst. Rübel* **3**:334–354.

Sauer, C. O. (1952). *Agricultural origins and dispersals*. American Geographical Soc., New York.

Simpson, G. G. (1951). *Horses*. Oxford Univ. Press.

Simpson, G. G. (1965) *The geography of evolution*. Chilton, Philadelphia.

Slobodkin, L. B. (1960). Ecological energy relationships at the population level. *Amer. Naturalist* **94**:213–236.

Summerhayes, V. S. (1941). The effect of voles (*Microtus agrestis*) on vegetation. *J. Ecol.* **29**:14–48.

Thorson, G. (1957). Bottom communities (sublittoral or shallow shelf). *Geol. Soc. Am. Mem.* **67**:461–534.

Vaurie, C. (1951). Adaptive differences between two sympatric species of nuthatches (*Sitta*). *Proc. X. Int. Orn. Cong.* **1950**:163–166.

Volterra, V. (1926). Variazione e fluttuazioni del numero d'individui in specie animali conviventi. *Mem. Accad. Naz. Lincei* **2**:31–113. (Abridged Translation in Chapman, R. N., 1931, *Animal Ecology*, McGraw-Hill, New York.

Waddington, C. H. (1957). *The strategy of the genes*. Allen and Unwin, London.

Walter, H., and H. Lieth (1960). *Klimadiagram Weltatlas*. G. Fischer, Jena.

Wright, S. (1931). Evolution in mendelian populations. *Genetics* **16**:97–159.

Wynne-Edwards, V. C. (1962). *Animal dispersion in relation to social behaviour*. Oliver and Boyd, Edinburgh.

Index